Python 程序设计
——从入门到实践
（中英双语版）

主　编　江业峰
副主编　张诗尧　高起跃　Alexander Galkin

北京理工大学出版社
BEIJING INSTITUTE OF TECHNOLOGY PRESS

内 容 简 介

本书详实地讲解了 Python 的基本概念、原理和使用方法，力求使学生打下扎实的程序设计基础，培养学生程序设计的能力。本书主要内容包括 Python 基础知识、Python 语法基础、控制结构、函数、列表与元组、字典与集合、文件、实战演练。本书采用循序渐进、深入浅出、通俗易懂的讲解方法，本着理论与实际相结合的原则，通过大量经典实例对 Python 进行了详细讲解，使程序设计语言的初学者能够掌握使用 Python 进行程序设计的技术和方法。

本书以 Python 编程基本技能训练为主线，内容完整，阐述准确，层次清楚。通过对本书的学习，学生将牢固掌握程序设计的基本技能，以适应信息时代对大学生的科学素质的要求。

本书适用于高等学校各专业程序设计基础教学，特别适合应用型本科、高职院校、中外合作办学的计算机及非计算机相关专业的学生使用，同时也是计算机等级考试备考的一本实用辅导书。

版权专有　侵权必究

图书在版编目（CIP）数据

Python 程序设计：从入门到实践：英、汉 / 江业峰主编. --北京：北京理工大学出版社，2024.11.
ISBN 978-7-5763-4040-2

Ⅰ．TP312.8

中国国家版本馆 CIP 数据核字第 2024RL4259 号

责任编辑：时京京	**文案编辑**：时京京	
责任校对：刘亚男	**责任印制**：李志强	

出版发行 / 北京理工大学出版社有限责任公司
社　　址 / 北京市丰台区四合庄路 6 号
邮　　编 / 100070
电　　话 /（010）68914026（教材售后服务热线）
　　　　　　（010）63726648（课件资源服务热线）
网　　址 / http://www.bitpress.com.cn
版 印 次 / 2024 年 11 月第 1 版第 1 次印刷
印　　刷 / 三河市天利华印刷装订有限公司
开　　本 / 787 mm×1092 mm　1/16
印　　张 / 16.75
字　　数 / 418 千字
定　　价 / 89.00 元

图书出现印装质量问题，请拨打售后服务热线，负责调换

前　言

　　Python 是一种在国际上非常流行的计算机程序设计语言，广泛应用于大数据、人工智能、物联网、云计算等领域。通过对 Python 这门语言的学习，学生可以掌握一门可直接用于求解复杂专业问题的编程语言，提高利用计算机解决问题的能力，从而具备在这个智能时代从事数据处理、人工智能等工作的基本能力。

　　本书在详细阐述程序设计基本概念、原理和方法的基础上，采用循序渐进、深入浅出、通俗易懂的讲解方法，本着理论与实际相结合的原则，通过大量经典实例重点讲解了 Python 的概念、规则和使用方法，便于程序设计语言的初学者能够在建立正确程序设计理念的前提下，掌握利用 Python 进行程序设计的技术和方法。全书共分为 8 章，主要内容包括：Python 基础知识、Python 语法基础、控制结构、函数、列表与元组、字典与集合、文件、实战演练。本书具有以下特色：一是中英双语编排，方便各高校中外合作办学专业或其他有双语教学需要的专业使用；二是语言表述简洁，通俗易懂，内容循序渐进、深入浅出，适合零基础学生和 Python 自学者使用；三是配套资源丰富，本书所有例题和课后习题均提供完整的 Python 参考代码，且代码注释均用英文表示。

　　本书使用大量丰富多彩的应用程序实例，讲解最实用的方法和技巧，以提高学生的计算机应用及编程能力，为后续工科专业课的学习奠定编程基础。通过 Python 程序设计这门课程的学习，学生能运用程序设计的基础知识和基本思想与方法，掌握高级语言程序设计的基本理论和基本技能，具备使用计算机解决问题的分析和程序设计能力，为使用计算机解决专业中的复杂工程计算问题打好基础。

　　本书是中外合作办学项目的主讲教材，外方编者为俄罗斯利佩兹克国立技术大学教授 Alexander Galkin。Alexander Galkin 具有多年教研经验，擅长大数据分析、数学建模和机器学习等领域，曾主持俄罗斯国家基础研究基金，发表高质量论文多篇，参编教材多部。Alexander Galkin 的参编，有力推进了中外办学合作项目的对接。

　　本书第 1 章、第 2 章由张诗尧编写；第 3 章、第 4 章、第 5 章、第 6 章由江业峰编写；第 7 章、第 8 章由高起跃编写；全书的英文校对工作由江业峰和 Alexander Galkin 共同完成。

　　由于编者水平有限，书中难免存在一些缺点和不足，殷切希望广大读者批评指正。

编　者

2024 年 6 月

目 录
CONTENTS

第 1 章　Python 基础知识 ··· 1
Chapter 1　Basic Knowledge of Python ··· 1
　1.1　Python 简介 ·· 1
　1.1　Introduction to Python ·· 1
　　1.1.1　Python 程序设计语言的发展过程 ··· 1
　　1.1.1　Development Process of Python Programming Language ················· 1
　　1.1.2　Python 的特点与应用 ··· 2
　　1.1.2　Features and Applications of Python ··· 2
　　1.1.3　程序设计语言及程序 ··· 4
　　1.1.3　Programming Language and Program ··· 4
　1.2　Python 的安装与使用 ·· 6
　1.2　Installation and Use of Python ·· 6
　　1.2.1　Python 的安装 ·· 6
　　1.2.1　Installation of Python ·· 6
　　1.2.2　Python 的运行 ·· 8
　　1.2.2　Running Python ··· 8
　1.3　Python 的标准输入输出 ·· 11
　1.3　Python's Standard Input and Output ·· 11
　　1.3.1　input( )函数 ·· 11
　　1.3.1　input( ) Function ·· 11
　　1.3.2　print( )函数 ·· 13
　　1.3.2　print( ) Function ··· 13
　1.4　集成开发环境 PyCharm 介绍 ··· 15
　1.4　Introduction to Integrated Development Environment PyCharm ················ 15
　　1.4.1　PyCharm 的下载 ·· 15
　　1.4.1　Download PyCharm ·· 15
　　1.4.2　PyCharm 的安装 ·· 16
　　1.4.2　Installation of PyCharm ·· 16
　　1.4.3　PyCharm 的简单使用 ·· 18
　　1.4.3　Simple Use of PyCharm ·· 18

1

1.5 本章小结 ………………………………………………………………… 21
1.5 Summary of This Chapter ……………………………………………… 21
1.6 本章练习 ………………………………………………………………… 22
1.6 Exercise for This Chapter ………………………………………………… 22

第 2 章　Python 语法基础 …………………………………………………… 23
Chapter 2　Basic Syntax of Python ………………………………………… 23

2.1 关键字和标识符 ………………………………………………………… 23
2.1 Keywords and Identifiers ……………………………………………… 23

2.1.1 关键字 ……………………………………………………………… 23
2.1.1 Keywords …………………………………………………………… 23
2.1.2 标识符 ……………………………………………………………… 24
2.1.2 Identifiers …………………………………………………………… 24

2.2 对象和变量 ……………………………………………………………… 25
2.2 Objects and Variables …………………………………………………… 25

2.2.1 对象 ………………………………………………………………… 25
2.2.1 Objects ……………………………………………………………… 25
2.2.2 变量 ………………………………………………………………… 26
2.2.2 Variables …………………………………………………………… 26
2.2.3 赋值 ………………………………………………………………… 26
2.2.3 Assignment ………………………………………………………… 26
2.2.4 多变量同步赋值 …………………………………………………… 27
2.2.4 Multiple Variables Simultaneous Assignment …………………… 27
2.2.5 变量命名规范 ……………………………………………………… 28
2.2.5 Variable Naming Conventions …………………………………… 28

2.3 数值类型 ………………………………………………………………… 29
2.3 Numerical Types ………………………………………………………… 29

2.3.1 整数类型 …………………………………………………………… 29
2.3.1 Integer Type ………………………………………………………… 29
2.3.2 布尔类型 …………………………………………………………… 31
2.3.2 Boolean Type ……………………………………………………… 31
2.3.3 浮点数类型 ………………………………………………………… 32
2.3.3 Floating-point Type ………………………………………………… 32
2.3.4 复数 ………………………………………………………………… 33
2.3.4 Complex Number …………………………………………………… 33

2.4 字符串类型 ……………………………………………………………… 34
2.4 String Type ……………………………………………………………… 34

2.4.1 字符串创建 ………………………………………………………… 34
2.4.1 String Creation ……………………………………………………… 34

目 录
CONTENTS

   2.4.2 转义字符 ·· 35
   2.4.2 Escape Characters ·································· 35
  2.5 数据类型转换 ··· 36
  2.5 Data Type Conversion ···································· 36
  2.6 运算符及表达式 ··· 41
  2.6 Operators and Expressions ································ 41
   2.6.1 算术运算符和表达式 ·································· 42
   2.6.1 Arithmetic Operators and Expressions ················· 42
   2.6.2 复合赋值运算符和表达式 ······························ 45
   2.6.2 Compound Assignment Operators and Expressions ······· 45
   2.6.3 比较运算符和表达式 ·································· 46
   2.6.3 Comparison Operators and Expressions ················ 46
   2.6.4 身份运算符和表达式 ·································· 47
   2.6.4 Identity Operators and Expressions ··················· 47
   2.6.5 成员运算符和表达式 ·································· 48
   2.6.5 Membership Operators and Expressions ··············· 48
   2.6.6 逻辑运算符和表达式 ·································· 49
   2.6.6 Logical Operators and Expressions ···················· 49
   2.6.7 运算符优先级 ······································· 51
   2.6.7 Operator Precedence ······························· 51
  2.7 常用数学函数 ··· 53
  2.7 Common Mathematical Functions ·························· 53
  2.8 标准库使用 ··· 56
  2.8 Standard Library Usage ·································· 56
   2.8.1 库和模块使用方法 ···································· 57
   2.8.1 How to Use Libraries and Modules ···················· 57
   2.8.2 math 模块 ·· 59
   2.8.2 math Module ······································ 59
   2.8.3 string 模块 ·· 60
   2.8.3 string Module ····································· 60
   2.8.4 random 模块 ······································ 61
   2.8.4 random Module ··································· 61
  2.9 字符串操作 ··· 64
  2.9 String Operations ······································· 64
   2.9.1 索引 ·· 64
   2.9.1 Index ·· 64
   2.9.2 切片 ·· 65
   2.9.2 Slice ··· 65
   2.9.3 拼接与重复 ·· 67

2.9.3　Concatenation and Repetition ……………………………………………… 67
　　　2.9.4　常用字符串处理方法 …………………………………………………… 71
　　　2.9.4　Common String Processing Methods …………………………………… 71
　2.10　本章小结 …………………………………………………………………………… 73
　2.10　Summary of This Chapter …………………………………………………………… 73
　2.11　本章练习 …………………………………………………………………………… 74
　2.11　Exercise for This Chapter …………………………………………………………… 74

第3章　控制结构 ………………………………………………………………………… 76
Chapter 3　Control Structure ……………………………………………………………… 76

　3.1　顺序结构 …………………………………………………………………………… 76
　3.1　Sequential Structure ………………………………………………………………… 76
　3.2　赋值语句 …………………………………………………………………………… 78
　3.2　Assignment Statement ……………………………………………………………… 78
　3.3　分支结构 …………………………………………………………………………… 80
　3.3　Conditional Structure ………………………………………………………………… 80
　　　3.3.1　单分支结构 ……………………………………………………………… 80
　　　3.3.1　Single Branch Structure ………………………………………………… 80
　　　3.3.2　缩进 ……………………………………………………………………… 81
　　　3.3.2　Indentation ……………………………………………………………… 81
　　　3.3.3　双分支结构 ……………………………………………………………… 82
　　　3.3.3　Double Branch Structure ………………………………………………… 82
　　　3.3.4　多分支结构 ……………………………………………………………… 85
　　　3.3.4　Multi-branch Structure …………………………………………………… 85
　　　3.3.5　分支结构嵌套 …………………………………………………………… 87
　　　3.3.5　Nested Conditional Structure …………………………………………… 87
　3.4　循环结构 …………………………………………………………………………… 89
　3.4　Loop Structure ……………………………………………………………………… 89
　　　3.4.1　range( )函数 ……………………………………………………………… 89
　　　3.4.1　range( ) Function ………………………………………………………… 89
　　　3.4.2　for循环 …………………………………………………………………… 91
　　　3.4.2　for Loop ………………………………………………………………… 91
　　　3.4.3　while循环 ………………………………………………………………… 95
　　　3.4.3　while Loop ……………………………………………………………… 95
　　　3.4.4　break语句 ………………………………………………………………… 100
　　　3.4.4　break Statement ………………………………………………………… 100
　　　3.4.5　continue语句 ……………………………………………………………… 102
　　　3.4.5　continue Statement ……………………………………………………… 102
　　　3.4.6　pass语句 ………………………………………………………………… 103
　　　3.4.6　pass Statement ………………………………………………………… 103

3.4.7 循环嵌套 ·················· 103
3.4.7 Nested Loop ·················· 103
3.4.8 循环中的 else 子句 ·················· 105
3.4.8 The else Clause within a Loop ·················· 105
3.5 异常处理 ·················· 107
3.5 Exception Handling ·················· 107
3.6 本章小结 ·················· 110
3.6 Summary of This Chapter ·················· 110
3.7 本章练习 ·················· 112
3.7 Exercise for This Chapter ·················· 112

第 4 章 函数 ·················· 114
Chapter 4 Functions ·················· 114

4.1 普通函数 ·················· 115
4.1 Regular Functions ·················· 115
    4.1.1 函数的定义 ·················· 115
    4.1.1 Function Definition ·················· 115
    4.1.2 函数的调用 ·················· 116
    4.1.2 Function Call ·················· 116
    4.1.3 函数的返回值 ·················· 118
    4.1.3 Return Value of a Function ·················· 118
    4.1.4 扩展例题 ·················· 119
    4.1.4 More Examples ·················· 119
4.2 函数的参数 ·················· 120
4.2 Function Parameters ·················· 120
4.3 函数的参数传递 ·················· 121
4.3 Parameter Passing in Functions ·················· 121
    4.3.1 位置传递 ·················· 121
    4.3.1 Positional Passing ·················· 121
    4.3.2 关键字传递 ·················· 122
    4.3.2 Keyword Passing ·················· 122
    4.3.3 默认值传递 ·················· 123
    4.3.3 Default Value Passing ·················· 123
    4.3.4 可变长参数传递 ·················· 124
    4.3.4 Variable-length Parameter Passing ·················· 124
    4.3.5 解包裹传递 ·················· 125
    4.3.5 Unpacking Passing ·················· 125
4.4 变量的作用域 ·················· 126
4.4 Scope of Variables ·················· 126
    4.4.1 局部变量 ·················· 126

4.4.1　Local Variables ……………………………………………… 126
　　　4.4.2　全局变量 …………………………………………………… 127
　　　4.4.2　Global Variables …………………………………………… 127
4.5　匿名函数 ………………………………………………………… 128
4.5　Anonymous Functions ………………………………………… 128
4.6　递归函数 ………………………………………………………… 129
4.6　Recursive Functions …………………………………………… 129
4.7　内置函数 ………………………………………………………… 131
4.7　Built-in Functions ……………………………………………… 131
　　　4.7.1　转换函数 …………………………………………………… 132
　　　4.7.1　Conversion Functions ……………………………………… 132
　　　4.7.2　序列操作函数 ……………………………………………… 134
　　　4.7.2　Sequence Operation Functions …………………………… 134
　　　4.7.3　help ( ) 函数 ………………………………………………… 135
　　　4.7.3　help ( ) Function …………………………………………… 135
4.8　本章小结 ………………………………………………………… 136
4.8　Summary of This Chapter ……………………………………… 136
4.9　本章练习 ………………………………………………………… 137
4.9　Exercise for This Chapter ……………………………………… 137

第 5 章　列表与元组 ……………………………………………… 139
Chapter 5　List and Tuple ………………………………………… 139

5.1　列表 ……………………………………………………………… 139
5.1　List ……………………………………………………………… 139
　　　5.1.1　列表的创建 ………………………………………………… 140
　　　5.1.1　Creating Lists ……………………………………………… 140
　　　5.1.2　元素的索引及访问 ………………………………………… 141
　　　5.1.2　Indexing and Accessing Elements ………………………… 141
　　　5.1.3　切片操作 …………………………………………………… 142
　　　5.1.3　Slicing Operation …………………………………………… 142
5.2　操作列表元素 …………………………………………………… 144
5.2　Manipulate List Elements ……………………………………… 144
　　　5.2.1　增加列表元素 ……………………………………………… 144
　　　5.2.1　Adding Elements to a List ………………………………… 144
　　　5.2.2　删除列表元素 ……………………………………………… 145
　　　5.2.2　Delete List Elements ……………………………………… 145
　　　5.2.3　修改列表元素 ……………………………………………… 148
　　　5.2.3　Modify List Elements ……………………………………… 148
　　　5.2.4　列表排序操作 ……………………………………………… 149
　　　5.2.4　List Sorting Operation ……………………………………… 149

目录 CONTENTS

  5.2.5　列表推导式 ·· 152
  5.2.5　List Comprehension ·· 152
 5.3　元组 ·· 153
 5.3　Tuple ·· 153
  5.3.1　元组的创建 ··· 153
  5.3.1　Creating Tuples ·· 153
  5.3.2　序列通用操作 ·· 154
  5.3.2　Common Sequence Operations ·· 154
 5.4　序列综合应用 ·· 158
 5.4　The Application of Sequence ·· 158
 5.5　本章小结 ··· 161
 5.5　Summary of This Chapter ·· 161
 5.6　本章练习 ··· 162
 5.6　Exercise for This Chapter ··· 162

第6章　字典与集合 ··· 165
Chapter 6　Dictionaries and Sets ··· 165

 6.1　字典的定义 ·· 165
 6.1　Definition of Dictionaries ··· 165
 6.2　字典的创建 ·· 166
 6.2　Creating Dictionaries ·· 166
  6.2.1　使用大括号创建字典 ··· 166
  6.2.1　Creating Dictionaries Using Curly Braces ······························· 166
  6.2.2　使用dict( )函数创建字典 ·· 166
  6.2.2　Creating Dictionaries Using the dict( ) Function ······················· 166
  6.2.3　使用map( )函数创建字典 ··· 168
  6.2.3　Creating Dictionaries Using the map( ) Function ······················ 168
  6.2.4　使用推导式创建字典 ··· 168
  6.2.4　Creating Dictionaries Using Comprehensions ··························· 168
  6.2.5　使用collections模块中的类创建字典 ····································· 169
  6.2.5　Creating Dictionaries Using Classes in the Collections Module ····· 169
 6.3　字典的访问 ·· 171
 6.3　Accessing Dictionaries ··· 171
  6.3.1　使用键访问 ··· 171
  6.3.1　Accessing by Keys ·· 171
  6.3.2　访问全部的键和值 ··· 172
  6.3.2　Retrieve All Keys and Values ··· 172
 6.4　字典元素的添加、修改和删除 ··· 173
 6.4　Adding, Modifying, and Deleting Dictionary Elements ····················· 173
 6.5　字典的更新与排序 ··· 176

6.5 Updating and Sorting the Dictionary ······ 176
6.6 字典拓展示例 ······ 177
6.6 Dictionary Expansion Examples ······ 177
6.7 集合的定义 ······ 179
6.7 Definition of a Set ······ 179
6.8 集合的创建 ······ 179
6.8 Creating a Set ······ 179
   6.8.1 使用大括号和set()函数创建集合 ······ 179
   6.8.1 Use Braces and the set() Function to Create Sets ······ 179
   6.8.2 使用推导式创建集合 ······ 181
   6.8.2 Use Set Comprehension to Create Sets ······ 181
6.9 集合元素的添加和删除 ······ 182
6.9 Adding and Removing Elements in a Set ······ 182
6.10 集合的运算 ······ 183
6.10 Set Operations ······ 183
6.11 集合拓展示例 ······ 184
6.11 Set Expansion Examples ······ 184
6.12 本章小结 ······ 186
6.12 Summary of This Chapter ······ 186
6.13 本章练习 ······ 186
6.13 Exercise for This Chapter ······ 186

第7章 文件 ······ 188
Chapter 7 File ······ 188

7.1 文件概述 ······ 188
7.1 Overview of Files ······ 188
   7.1.1 文件与文件类型 ······ 189
   7.1.1 Files and File Types ······ 189
   7.1.2 文件存储路径 ······ 190
   7.1.2 File Storage Path ······ 190
   7.1.3 文件处理流程 ······ 191
   7.1.3 File Processing Flow ······ 191
7.2 文件的打开与关闭 ······ 192
7.2 Opening and Closing Files ······ 192
   7.2.1 文件的打开操作 ······ 192
   7.2.1 File Opening Operation ······ 192
   7.2.2 文件的关闭操作 ······ 195
   7.2.2 File Closing Operation ······ 195
   7.2.3 文件的基本操作实例 ······ 197
   7.2.3 Basic File Operation Examples ······ 197

目　录 CONTENTS

7.3　文件的基本操作 ·· 200
7.3　Basic Operations of Files ·· 200
　　7.3.1　文件的读写操作 ··· 200
　　7.3.1　Reading and Writing Operations on Files ···································· 200
　　7.3.2　文件的定位操作 ··· 205
　　7.3.2　File Positioning Operations ·· 205
　　7.3.3　文件的其他操作 ··· 207
　　7.3.3　Other Operations on Files ··· 207
7.4　文件的综合应用 ·· 209
7.4　Integrated Application of Files ··· 209
7.5　本章小结 ·· 212
7.5　Summary of This Chapter ·· 212
7.6　本章练习 ·· 214
7.6　Exercise for This Chapter ·· 214

第8章　实战演练 ·· 216
Chapter 8　Practical Exercises ·· 216
8.1　学生成绩管理系统 ··· 216
8.1　Student Performance Management System ·· 216
8.2　学生学业成绩统计 ··· 231
8.2　Student Academic Performance Statistics ··· 231
8.3　绘制词云 ·· 237
8.3　Generating Word Cloud ··· 237
8.4　绘制七段管 ·· 241
8.4　Drawing a Seven-segment Display ··· 241
8.5　使用scipy库进行图像处理 ·· 245
8.5　Using the scipy Library for Image Processing ···································· 245
8.6　本章小结 ·· 250
8.6　Summary of This Chapter ·· 250
8.7　本章练习 ·· 251
8.7　Exercise for This Chapter ·· 251

参考文献 ·· 253

第 1 章　Python 基础知识
Chapter 1　Basic Knowledge of Python

1.1　Python 简介
1.1　Introduction to Python

Python 由荷兰数学和计算机科学国家研究学会的吉多·范罗苏姆于 1990 年代初设计。Python 自诞生以来就备受瞩目，它不仅提供了高级数据结构，还让编程变得简单而富有成效。

Python 解释器的设计使其易于扩展，开发者可以使用 C 或 C++（或其他可与 C 交互的语言）来扩展新的功能和数据类型。在需要定制化的软件中，Python 也常被用作扩展程序语言。

Python 在实际中应用广泛，包括人工智能、网络爬虫、系统自动化和游戏开发等。腾讯、网易和搜狐等大型企业也广泛使用 Python 进行开发。

Python, designed in the early 1990s by Guido van Rossum of the Netherlands' National Research Institute for Mathematics and Computer Science. Python has been highly anticipated since its creation, it not only provides advanced data structures, but also makes programming simple and productive.

The design of the Python interpreter makes it easy to extend, allowing developers to use C or C++(or other languages can interact with C) to add new features and data types. Python is also commonly used as an extension programming language in customized software.

Python is widely used in practical applications, including artificial intelligence, web scraping, system automation and game development. Large enterprises such as Tencent, NetEase and SOHU extensively employ Python for development purposes.

1.1.1　Python 程序设计语言的发展过程
1.1.1　Development Process of Python Programming Language

1989 年：吉多·范罗苏姆开始创建 Python。

1991 年：Python 的第一个公

In 1989: Guido van Rossum started creating the Python.

In 1991:The first public version of Python(0.9.0)

开版本(0.9.0)发布。

1994 年：Python 1.0 发布，加入了 lambda 表达式和函数式编程元素。

2000 年：Python 2.0 发布，引入了列表推导式和垃圾回收器等功能。

2008 年：Python 3.0 发布，这是一次重大的更新，引入了许多不兼容的变化，以解决 Python 2.x 版本中的一些设计缺陷。

2014 年：Python 3.4 发布，引入了协程和异步编程的支持，为编写高效的异步代码提供了便利。

2017 年：Python 3.6 发布，这个版本引入了异步生成器和"f-strings"等新特性。

2020 年：Python 3.9 发布，为 Python 引入了许多新功能，包括新的解释器功能和更好的性能优化。

Python 程序设计语言以简洁的语法和易读性为特点，支持面向对象和面向过程编程。随着版本的更新，Python 的功能不断增强，逐渐成为多数平台上编写脚本和快速开发应用的编程语言。如今，Python 的应用领域不断扩大，包括数据科学、人工智能、网络爬虫、系统自动化等。

was released.

In 1994: Python 1.0 was released, introducing lambda expressions and functional programming elements.

In 2000: Python 2.0 was released, introducing features such as list comprehensions and garbage collector.

In 2008: Python 3.0 was released, a major update that introducing many incompatible changes to address design flaws in Python 2.x versions.

In 2014: Python 3.4 was released, introducing support for coroutines and asynchronous programming, making it convenient to write efficient asynchronous code.

In 2017: Python 3.6 was released, introducing new features such as asynchronous generators and "f-strings".

In 2020: Python 3.9 was released, introducing many new features to Python, including new interpreter features and better performance optimizations.

The Python programming language is characterized by its concise syntax and readability, supports object-oriented and procedural programming. With the updates of its version, Python's functionality has been continuously enhanced, and it has gradually become a programming language for scripting and rapid application development on most platforms. Nowadays, the application areas of Python are constantly expanding, including data science, artificial intelligence, web crawling, system automation, etc.

1.1.2 Python 的特点与应用
1.1.2 Features and Applications of Python

Python 因易学且强大而受程序员喜爱，其高效数据结构、面向对象编程及简洁语法等特点，使其成为多个领域快速开发的理想语言。随着大数据、人工智能的发展，Python 日益受到重视。

Python is loved by programmers for its ease of learning and powerful capabilities. Its efficient data structures, object-oriented programming, and concise syntax make it an ideal language for rapid development in multiple fields. With the development of big data and artificial intelligence, Python is receiving attention increasingly.

第1章 Python基础知识
Chapter 1　Basic Knowledge of Python

　　Python 是一种高级编程语言，它具有许多优点，使得它成为开发人员和科学家的首选语言之一。以下是 Python 的一些主要优点。

　　（1）简单易学：Python 的语法清晰简洁，易于理解和学习。这使得 Python 成为初学者的理想选择，同时也使得经验丰富的开发人员能够更高效地编写代码。

　　（2）可移植性强：Python 可以在多种操作系统上运行，包括 Windows、Linux 和 macOS 等。这使得 Python 成为一种非常灵活的语言，可以轻松地在不同的平台上开发和部署应用程序。

　　（3）开源免费：Python 是开源的，这意味着任何人都可以查看、使用和修改其源代码。此外，Python 还有大量的免费库和工具可供使用，这降低了开发成本。

　　（4）丰富的库和框架：Python 拥有庞大的标准库和第三方库，如 Web 开发、数据分析、机器学习、科学计算等。

　　Python 是一种功能强大的编程语言，但它也有一些不足之处。以下是 Python 的一些主要缺点。

　　（1）执行速度慢：与其他编译型语言如 C++或 Java 相比，Python 的执行速度相对较慢。这是因为它是一种解释型语言，每次运行代码时都需要动态解释和执行，在某些性能敏感的应用场景中，Python 可能不是最佳选择。

　　（2）版本不兼容：Python 2 和 Python 3 之间的不兼容问题在过去曾经是一个重大挑战，尽管现在 Python 2 已经过时且不再受支持，但一些旧的库和工具可能仍然需要

　　Python is a high-level programming language, it has many advantages that make it one of the preferred languages for developers and scientists. Here are some of the main advantages of Python.

　　(1) Easy to learn: Python's syntax is clear and concise, making it easy to understand and learn. This makes Python an ideal choice for beginners, while also enabling experienced developers to write code more efficiently.

　　(2)High portability: Python can run on multiple operating systems, including Windows, Linux, macOS, etc. This makes Python a very flexible language that can easily develop and deploy applications on different platforms.

　　(3) Open source and free: Python is open source, which means anyone can view, use, and modify its source code. In addition, Python has a large number of free libraries and tools, which reduces development costs.

　　(4)Rich libraries and frameworks: Python has a large standard library and third-party libraries, such as Web development, data analysis, machine learning, scientific computing, etc.

　　Python is a powerful programming language, but it also has some shortcomings. Here are some of the main disadvantages of Python.

　　(1) Slow execution speed: Compared to other compiled languages such as C++ or Java, Python's execution speed is relatively slow. This is because it is an interpreted language that requires dynamic interpretation and execution each time the code runs. Python may not be the best choice in certain performancesensitive application scenarios.

　　(2)Version incompatibility: incompatibility between Python 2 and Python 3 used to be a major challenge in the past. Although Python 2 is now outdated and no longer supported, some legacy libraries and tools may still require a Python 2 environment.

Python 2 环境。

尽管 Python 有这些不足，但它仍然是一种非常流行和有用的编程语言，特别适用于快速开发、原型设计、数据分析、机器学习、Web开发等多个领域。在选择编程语言时，重要的是要根据项目的具体需求和团队的技能来权衡各种因素。

Despite these shortcomings, Python remains a very popular and useful programming language, particularly well-suited for rapid development, prototyping, data analysis, machine learning, web development and many other fields. When choosing a programming language, it is important to consider the specific needs of the project and the skills of the team to weigh various factors.

1.1.3 程序设计语言及程序
1.1.3 Programming Language and Program

程序设计语言是人与计算机交流的工具，用于指定计算机执行的任务。它是一套标准化的语法、语义和规则的集合，使程序员能够编写指令，控制计算机硬件和软件资源，解决各种计算问题。程序则是使用某种程序设计语言编写的一系列指令的集合，用于指导计算机完成特定任务或解决特定问题。程序可以被看作是一种软件，它可以是简单的几个指令，也可以是复杂的系统软件或应用软件。

Programming language is a tool for human-computer communication, used to specify tasks for the computer to perform. It is a set of standardized syntax, semantics and rules that enable programmers to write instructions, control computer hardware and software resources, solve various computational problems. A program is a set of instructions written in a programming language that instructs a computer to complete specific tasks or solve specific problems. A program can be viewed as a type of software, which can be as simple as a few instructions or as complex as system software or application software.

计算机程序设计语言的发展经历了机器语言、汇编语言和高级语言 3 个阶段，其形式如表 1-1 所示。

The development of computer programming languages has gone through three stages: machine language, assembly language and high-level language, as shown in Table 1-1.

表 1-1 3 种语言求解"5-3"的比较
Table 1-1 Comparison of Three Languages for Solving the "5-3"

机器语言 Machine Language	汇编语言 Assembly Language	高级语言 High-level Language
LOAD ACCUMULATOR, #5 SUBTRACT ACCUMULATOR, #3 STORERESULT, MEMORY_ADDRESS	section. data num1 dw 5 num2 dw 3 result dw 0 section. text global _start _start: mov ax, [num1] sub ax, [num2] mov [result], ax	num1 = 5 num2 = 3 result = num1 − num2 print(result)

机器语言：直接对应于计算机的硬件指令集，由二进制代码组成，难以阅读和编写。

汇编语言：用助记符代替机器语言中的操作码，易于记忆和书写，但仍需通过汇编器转换成机器语言。

面向过程的语言：如 C、Fortran 等，强调解决问题的过程或步骤。

面向对象的语言：如 Java、C++、Python 等，以对象为基础，强调数据和操作的封装。

脚本语言：如 JavaScript、Shell 等，通常用于自动化任务或网站开发。

程序结构：

(1)顺序结构：程序按照代码的顺序从上到下执行。

(2)分支结构：根据条件判断的结果，程序可以选择不同的执行路径，如 if-else 语句、switch 语句等。

(3)循环结构：程序可以重复执行一段代码，直到满足退出条件，如 for 循环、while 循环等。

程序开发过程：

(1)需求分析：确定程序需要解决什么问题，收集用户需求。

(2)设计：规划程序的结构和功能，选择合适的数据结构和算法。

(3)编码：使用选定的程序设计语言编写代码。

(4)测试：检查程序是否按照预期工作，修复发现的错误。

(5)部署与维护：将程序部署到生产环境，并持续监控和维护。

以下是一个简单的 Python 程序

Machine language: It directly corresponds to the hardware instruction set of a computer and consists of binary code, making it difficult to read and write.

Assembly language: It uses mnemonics instead of machine language opcodes, making it easier to remember and write, but it still requires conversion to machine language through an assembler.

Procedural languages: Such as C and Fortran, emphasize the process or steps of solving problems.

Object-oriented languages: Such as Java, C++, Python, etc, are based on objects and emphasize the encapsulation of data and operations.

Scripting languages: Such as JavaScript, Shell, etc, are often used for automating tasks or website development.

Program structure:

(1) Sequence structure: The program is executed from top to bottom in the order of the code.

(2) Conditional structure: Based on the results of conditional judgment, the program can choose different execution paths, such as if-else statements, switch statements, and so on.

(3) Looping structure: Programs can repeatedly execute a block of code until a condition is met, such as for loops, while loops, and so on.

Program development process:

(1)Requirement analysis: Determine what problems the program needs to solve and collect user requirements.

(2)Design: Plan the structure and function of the program, and choose appropriate data structures and algorithms.

(3)Coding: Write code using the selected programming language.

(4)Testing: Check whether the program works as expected and fix any errors found.

(5)Deployment and maintenance: Deploy the program to the production environment and continuously monitor and maintain it.

The following is a simple Python program example

示例，用于计算并输出两个数的和。 | for calculating and outputting the sum of two numbers.

```
num1 = float(input("Please input the first number: "))
num2 = float(input("Please input the second number: "))
sum = num1+num2
print("The sum of two numbers is: ", sum)
```

1.2 Python 的安装与使用
1.2 Installation and Use of Python

1.2.1 Python 的安装
1.2.1 Installation of Python

Python 主要有两个版本系列：Python 2.x 和 Python 3.x。这两个版本系列之间存在一些不兼容的差异，因此，在学习或开发时，需要明确所使用的版本。本书基于广泛使用的 Python 3.9 版本进行编写，这个版本提供了丰富的功能和良好的性能，是学习和开发应用的理想选择。

Python 是一种跨平台语言，这意味着它可以在多种操作系统上运行，包括 Windows、macOS 以及各种 Linux/UNIX 发行版。因此，无论是哪种操作系统，都可以轻松地安装和使用 Python。

以 Windows 系统为例，简述 Python IDLE(集成开发环境)安装步骤。

（1）访问 Python 的官方网站，在下载页面选择适合 Windows 系统的 Python 安装包进行下载。请注意选择与计算机系统相匹配的版本（32 位或 64 位）。

（2）下载完成后，双击安装包开始安装，安装界面如图 1-1 所示。在安装过程中，可以选择默认

Python is divided into two main versions: Python 2.x and Python 3.x. There are some incompatible differences between these two versions, so it is important to be clear about which version to use when learn or develop. This book is based on the widely used Python 3.9, which provides rich features and good performance. It is an ideal choice for Python learning and developing Python applications.

Python is a cross-platform language, which means it can run on multiple operating systems, including Windows, macOS, and various Linux/UNIX distributions. Therefore, no matter which operating system it is, one can easily install and use Python.

Take the Windows system as an example to briefly describe the installation steps of Python IDLE.

(1)Visit the official website of Python, and select the Python installation package suitable for the Windows system from the download page to download. Please note to select the version that matches the computer system(32-bit or 64-bit).

(2)After the download is complete, doubleclick on the installation package to begin the installation process. The installation interface is shown in Figure 1-1. During installation,

第1章　Python基础知识
Chapter 1　Basic Knowledge of Python

安装或自定义安装。如果选择自定义安装，请确保勾选 Add Python 3.9 to PATH 选项，程序可在命令提示符中直接运行。

one can choose between the default installation and a custom installation. If one opts for a custom installation, make sure to check the Add Python 3.9 to PATH option, which will allow Python to be run directly from the command prompt.

图 1-1　Python 3.9 安装界面

Figure 1-1　Python 3.9 Installation Interface

安装过程中，还可以选择安装其他组件，如 pip（Python 包管工具），根据需求进行选择。

Python 3.9 安装完成界面如图 1-2 所示，Python 和 IDLE 应该已经成功安装在计算机上。可以在开始菜单中找到 Python 文件夹，并打开 IDLE（Python GUI）来启动 IDLE。

During the installation, one can also choose to install additional components such as pip(Python package management tools). Select them based on the requirements.

Python 3.9 installation completion interface is shown in Figure 1-2, Python and IDLE should be successfully installed on the computer. One can find the Python folder in the start menu and launch IDLE by opening IDLE(Python GUI).

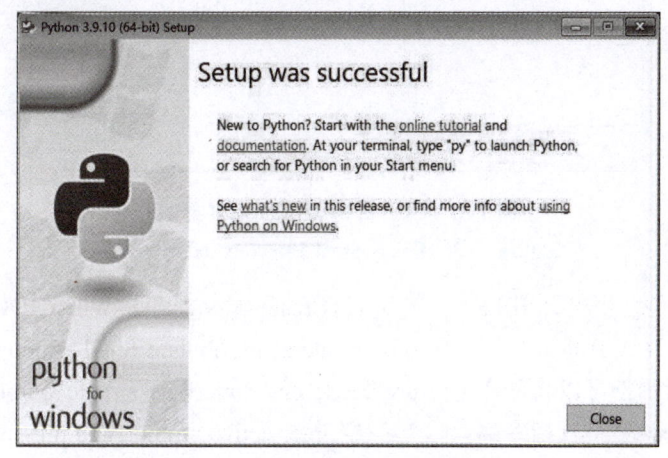

图 1-2　Python 3.9 安装完成界面

Figure 1-2　Python 3.9 Installation Completion Interface

7

1.2.2 Python 的运行
1.2.2 Running Python

通过上一节的安装学习之后，接下来使用 Python 的 IDLE 编辑器输出"hello world"字符串，从运行到输出结果，分步骤进行介绍。

（1）打开 IDLE 编辑器。首先，确保已经安装了 Python，并且 IDLE 也被同时安装（通常 Python 的默认安装会包含 IDLE）。

通过开始菜单找到并启动 IDLE。在 Windows 上，可以直接搜索"IDLE"或者在 Python 文件夹下找到它，启动 IDLE，进入 Python 的 IDLE Shell 界面，如图 1-3 所示。在 macOS 上，可以在应用程序/Python 文件夹下找到它。在 Linux 上，则可以通过终端输入 IDLE 来启动。

After learning the installation in the previous section, we will use Python's IDLE editor to output the string "hello world". We will introduce the steps from running to outputting the result.

(1)Open the IDLE editor. First, make sure one has Python installed and that IDLE is also installed(Typically, Python's default installation includes IDLE).

Next, find and launch IDLE from the start menu. On Windows, one can search for "IDLE" or find it under the Python folder, start IDLE as shown in Figure 1-3 to enter the IDLE Shell interface of Python. On macOS, one can find it under the Applications/Python folder. On Linux, one can launch it by entering IDLE in the terminal.

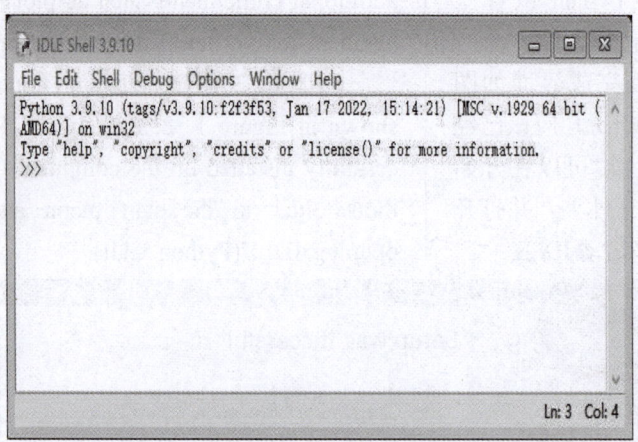

图 1-3　Python 的 IDLE Shell 界面
Figure 1-3　IDLE Shell Interface of Python

（2）创建新文件。当打开 IDLE 后，会看到一个交互式 Python Shell 窗口，如图 1-4 所示，可以将简单的输出代码写在">>>"交互指令后面，按〈Enter〉键，即可查看代码的运行结果。

但是，为了编写一个完整的程

(2)Create a new file. When IDLE opens, one will see an interactive Python Shell window, as shown in Figure 1-4, one can write simple output code behind the ">>>" inter-active prompt, and press ⟨Enter⟩ to view the output result in codes.

However, in order to write a complete program, we

第1章　Python基础知识
Chapter 1　Basic Knowledge of Python

序，需要创建一个新的文件。在 IDLE 的菜单栏上，选择 File→New File 命令（或者使用快捷键〈Ctrl+N〉）。这将打开一个新的文本编辑器窗口，可以在这里编写 Python 代码。

need to create a new file. On the menu bar of IDLE, select File → New File command (or use the shortcut 〈Ctrl+N〉). This will open a new text editor window where one can write Python code.

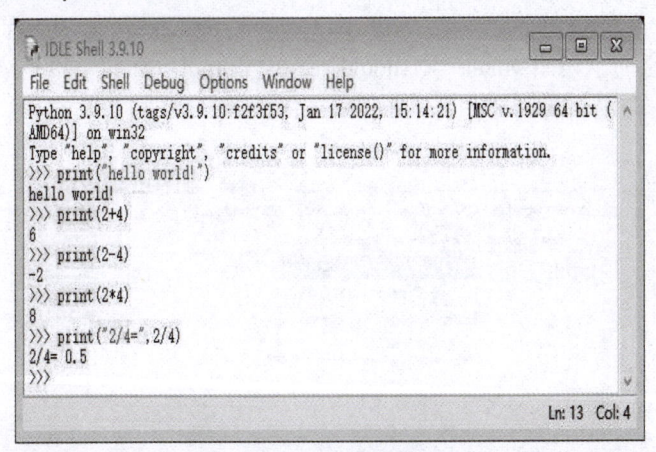

图 1-4　Python Shell 窗口
Figure 1-4　Python Shell Window

（3）编写代码。在新打开的文本编辑器窗口中，输入以下代码：print("hello world!")。这行代码是 Python 程序，它的作用是输出字符串"hello world!"，如图 1-5 所示。

(3) Write the code. In the newly opened text editor window, enter the following code: print("hello world!"). This line of code is a Python program, and its purpose is to output the string "hello world!", as shown in Figure 1-5.

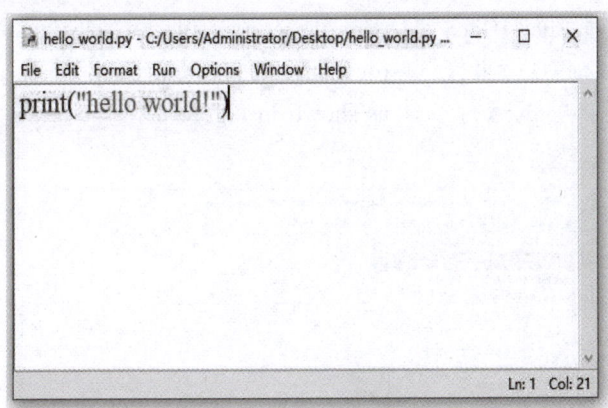

图 1-5　编写代码
Figure 1-5　Writing Code

（4）保存文件。在菜单栏上，选择 File→Save 命令（或者使用快捷键〈Ctrl+S〉）。在弹出的"保存"对话框中，选择一个位置来保存

(4) Save the file. On the menu bar, select File→Save command (or use the shortcut 〈Ctrl+S〉). In the pop-up "save" dialog, select a location to save the file, and give it a name, such as hello_world.py. Make sure the file ex-

9

文件，并给它起一个名字，比如 hello_world.py。确保文件扩展名是 .py，这是 Python 源代码文件的标准扩展名。

（5）运行代码。选择 Run Module 命令就会运行这个文件中的代码，如图 1-6 所示。这时会在 Python Shell 窗口中看到输出的字符串"hello world！"。

tension is .py, which is the standard extension for Python source code files.

(5) Run the code. Selecting the Run Module command will run the code in this file, as shown in Figure 1-6. One should see the output string "hello world！" in the Python Shell window.

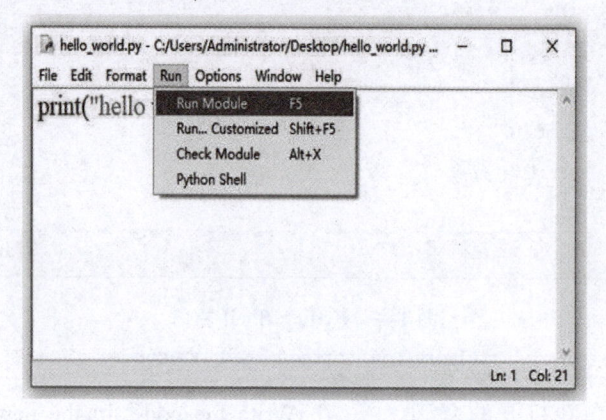

图 1-6 执行 hello_world.py 程序

Figure 1-6 Executing the hello_world.py Program

请注意，如果在步骤（5）中 Run Module 命令没有运行期望的文件，则可能需要先通过 File→Open 命令（或使用快捷键〈Ctrl+O〉）来打开正确的文件，然后再尝试运行，如图 1-7 所示。

Note that if the Run Module command does not run the file one expects in step (5), one may need to first open the correct file by selecting File→Open command (or using the shortcut 〈Ctrl+O〉), and then try to run it, as shown in Figure 1-7.

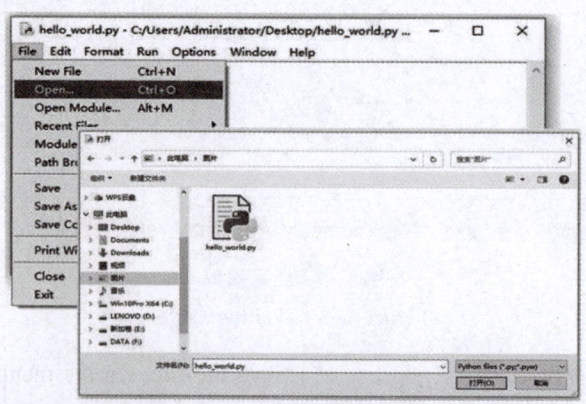

图 1-7 打开和执行 hello_world.py 文件

Figure 1-7 Opening and Executing the hello_world.py File

第1章　Python基础知识
Chapter 1　Basic Knowledge of Python

最后会在 Python Shell 窗口中看到输出的字符串"hello world!"，如图 1-8 所示。

Finally, one will see the output string "hello world!" in the Python Shell window, as shown in Figure 1-8.

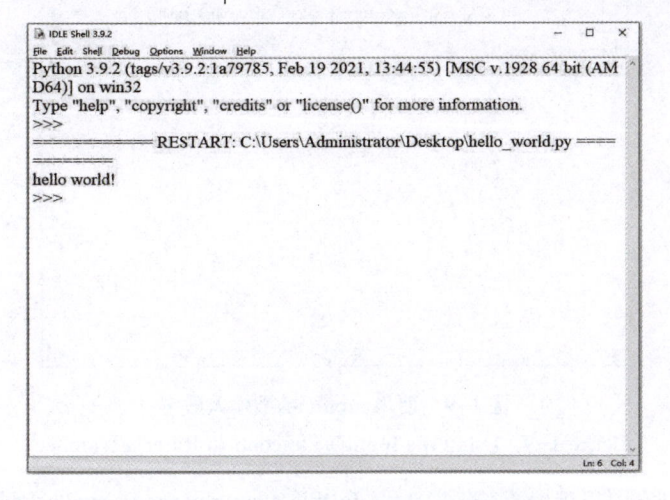

图 1-8　执行 hello_world.py 的结果

Figure 1-8　Output After Executing hello_world.py

1.3　Python 的标准输入输出
1.3　Python's Standard Input and Output

1.3.1　input()函数
1.3.1　input() Function

在 Python 中，input()函数用于从用户处获取输入。当 input()函数被调用时，程序会停止并等待用户在交互式编辑界面的命令行中输入一些文本或数字。

用户输入的内容将被作为字符串返回。下面是一个简单的例子。

In Python, the input() function is used to obtain input from the user. When input() function is called, the program will stop and wait for the user to enter text or numbers in the command line of the interactive editing interface.

The content entered by the user will be returned as a string. Here is a simple example.

```
number=input("Please enter a number:")
number=int(number)    #Convert a string to an integer
print("The double of the number you input is "+str(number*2))
```

运行结果如图 1-9 所示。

The running result is shown in Figure 1-9.

11

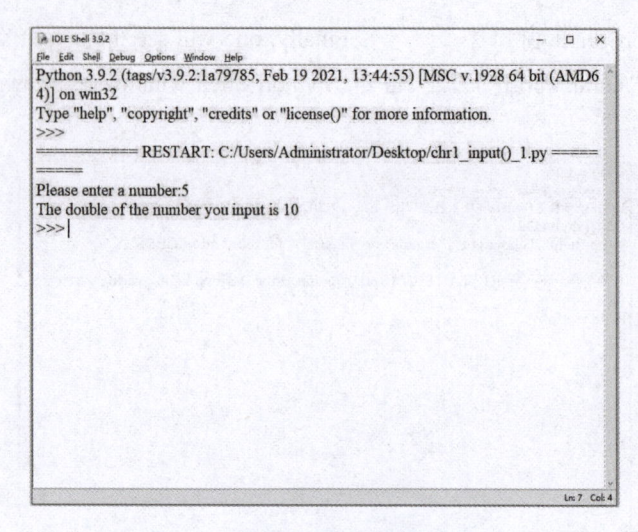

图 1-9 使用 input() 函数输入数字

Figure 1-9 Using the input() Function to Enter a Number

在这个例子中，用户被要求输入一个数字。输入被读取为字符串，然后使用 int() 函数将其转换为整数。之后，计算该数字的双倍，并使用 str() 函数将结果转换回字符串，以便可以将其与前面的文本连接并打印出来。

在 Python 中，input() 函数的语法非常简单。它接收一个可选的参数，通常是一个字符串，用作提示用户输入的消息。当用户输入内容后，input() 函数将返回一个包含用户输入的字符串。下面是 input() 函数的基本语法。

In this example, the user is asked to enter a number. The input is read as a string and converted to an integer using the int() function. Then, the number is doubled and the result is converted back to a string using the str() function so that it can be concatenated with the previous text and printed out.

In Python, the syntax of the input() function is very simple. It receives an optional parameter, usually a string, which is used as a message to prompt the user for input. When the user enters something, the input() function returns a string containing the user's input. Here is the basic syntax of the input() function.

```
input(prompt)
```

prompt：这是一个可选参数，表示要显示给用户的提示或消息，它可以是一个字符串，用作输入数据前的提示性消息，也可以不设置这个参数。下面是一个使用 input() 函数的例子。

prompt: This is an optional parameter that represents the prompt or message to display to the user. It can be a string used as a prompt message before entering data, or this parameter can be left unset. Here is an example using the input() function.

```
user_name=input("Please enter your name:")
print("Hello,"+user_name+"!")
```

第1章　Python基础知识
Chapter 1　Basic Knowledge of Python

运行结果如图 1-10 所示。

The running result is shown in Figure 1-10.

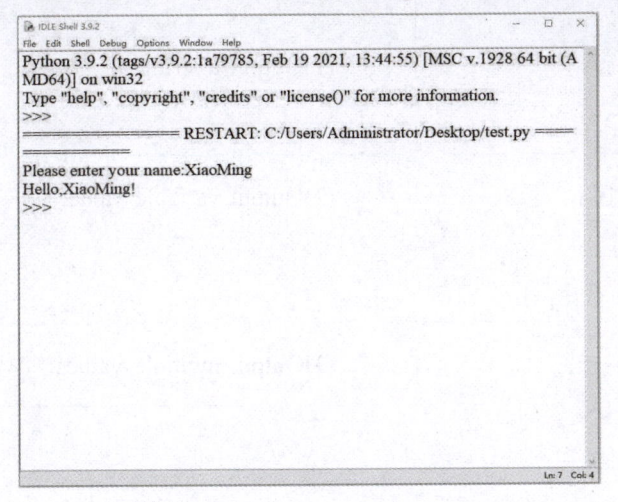

图 1-10　使用 input() 函数输入字符串

Figure 1-10　Using the input() Function to Enter a String

需要注意的是，无论用户输入的是什么类型的数据（数字、文本等），input() 函数都会将其当作字符串返回。如果需要将用户的输入转换为其他类型（如整数或浮点数），就需要使用相应的转换函数，如 int() 或 float() 等数值类型转换函数。

It's important to note that regardless of the type of data entered by the user(numbers, text, etc.), the input() function will always return it as a string. If one needs to convert the user's input to another type(such as an integer or a float), one will need to use the corresponding conversion function, such as int() or float(), for numeric type conversion.

1.3.2　print() 函数
1.3.2　print() Function

在 Python 中，print() 函数是一个内置函数，用于将运行结果输出到交互式编辑界面。这个函数非常基础且重要，因为它允许程序员输出程序的结果信息，一个 Python 程序可以没有输入函数但至少要有一个输出函数。

print() 函数可以输出多个参数，这些参数可以是字符串、数字等，它们会被依次输出到交互式编辑界面，并且默认以空格分隔。此外，print() 函数还可以接收一些可选参数来设置输出的格式。

In Python, the print() function is a built-in function that is used to output the running results to the interactive editing interface. This function is very basic and important, as it allows programmers to output program results. A Python program can have no input function but it must have at least one output function.

The print() function can output multiple parameters, such as strings, numbers, etc. They will be output to the interactive editing interface sequentially, and separated by spaces by default. In addition, the print() function can also receive some optional parameters to setup the output format.

13

下面是 print() 函数的一些基本用法。

(1) 输出字符串：

```
print("Hello, World!")
```

(2) 输出变量的值：

```
x=10
print("The value of x is:", x)
```

(3) 输出多个值：

```
a=5
b=10
print(a,b)
```

(4) 使用 sep 参数改变分隔符：

```
print("apple", "banana", "cherry", sep=",")
```

(5) 使用 end 参数改变行尾字符（默认是换行符"\n"）：

```
print("This is a line.", end=".")
print("This is another line.")
```

print() 函数还有一个特点是支持字符串格式化，即可以将变量的值嵌入字符串中。Python 3.6 引入了 f-string 作为一种更简洁的字符串格式化方法，但在 print() 函数中仍然可以使用传统的字符串格式化方法，如%操作符或 str.format() 方法。

例如，使用%操作符：

```
name="Alice"
age=30
print("My name is %s and I am %d years old." %(name, age))
```

或者使用 str.format() 方法：

```
print("My name is {} and I am {} years old.".format(name, age))
```

或者使用 f 前缀：

```
print(f"My name is {name} and I am {age} years old.")
```

Here are some basic usages of the print() function.

(1) Output a string:

(2) Output variable value:

(3) Output multiple values:

(4) Use the sep parameter to change the delimiter:

(5) Use the end parameter to change the line ending character (By default, it is the newline character "\n"):

Another feature of the print() function is its support for string formatting, which allows one to embed variable values into strings. Python 3.6 introduced f-string as a more concise string formatting method, but traditional string formatting methods such as the % operator or str.format() method can still be used in the print() function.

For example, use the % operator:

Or use the str.format() method:

Or use the f prefix:

第1章　Python基础知识
Chapter 1　Basic Knowledge of Python

这些格式化方法允许创建动态的输出内容，其中，变量的值可以在运行时被插入字符串中的指定位置。

These formatting methods allow one to create dynamic output content, where the value of a variable can be inserted into a specified position in the string at run time.

1.4　集成开发环境 PyCharm 介绍
1.4　Introduction to Integrated Development Environment PyCharm

PyCharm 是一种由 JetBrains 公司开发的 Python 集成开发环境，旨在提高开发者的工作效率。它提供了功能强大的工具，包括调试、语法高亮、项目管理、代码跳转、智能提示、自动完成、单元测试以及版本控制等，这些功能可以帮助开发者更快地编写代码、更容易地找到和修复错误、更快地测试代码以及部署应用程序。

PyCharm is a Python integrated development environment developed by JetBrains, which aims to improve developer productivity. It provides powerful tools, including debugging, syntax highlighting, project management, code jumping, intelligent prompts, auto completion, unit testing and version control. These features help developers write code faster, find and fix errors more easily, test code and deploy applications faster.

此外，PyCharm 还具备一些高级功能，比如支持 Django 框架下的专业 Web 开发，以及可视化接口和自动部署等。同时，PyCharm 也支持多种 Python 解释器，并可以与多种数据库和 Web 服务器进行集成。

In addition, PyCharm also has some advanced features, such as supporting professional Web development under the Django framework, as well as visual interfaces and automatic deployment. At the same time, PyCharm also supports multiple Python interpreters and can integrate with multiple databases and Web servers.

1.4.1　PyCharm 的下载
1.4.1　Download PyCharm

以下是 PyCharm 的下载步骤。

（1）打开浏览器，访问 JetBrains 的官方网站，也可以在搜索引擎中输入"PyCharm 下载"来找到相关的下载页面，如图 1-11 所示。

Here are the steps to download PyCharm.

(1) Open the browser and visit the official website of JetBrains. One can also enter "PyCharm download" in the search engine to find the relevant download page, as shown in Figure 1-11.

15

图 1-11　PyCharm 的官网首页

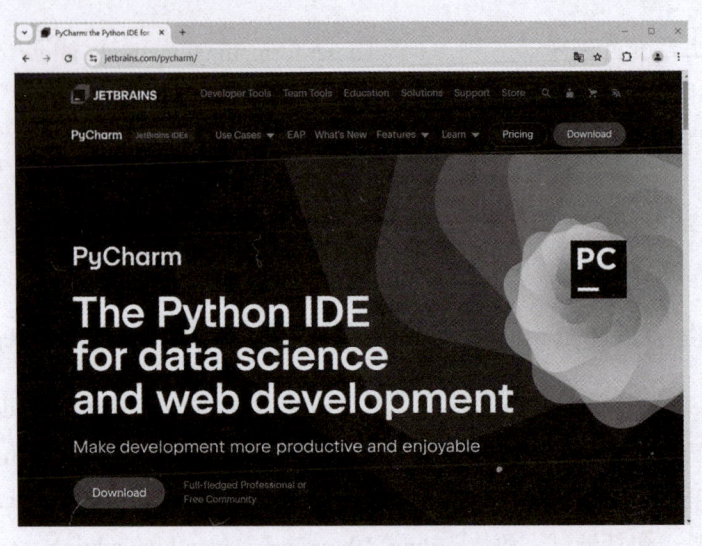

Figure 1-11　Home Page of Official PyCharm Website

（2）单击所选版本下方的"下载"按钮开始下载。下载文件通常是一个 .exe 可执行文件（Windows）或 .dmg 文件（macOS），如图 1-12 所示。

(2)Click the "Download" button below the selected version to start downloading. The download file is usually an .exe executable file(Windows) or a .dmg file(macOS), Figure 1-12 shows the PyCharm download interface.

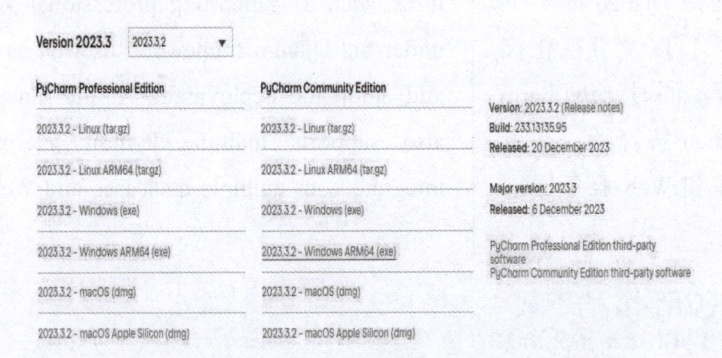

图 1-12　PyCharm 的下载界面

Figure 1-12　PyCharm Download Interface

1.4.2　PyCharm 的安装
1.4.2　Installation of PyCharm

下载完成后，找到下载的文件并双击运行。在 Windows 上，用户将看到一个安装向导，其将指导用户完成安装过程。

After the download is complete, locate the downloaded file and double-click to run it. In Windows, the user will see an installation wizard that guides the user through the installation process.

第1章　Python基础知识
Chapter 1　Basic Knowledge of Python

请按照安装向导的指示进行操作，包括同意许可协议、选择安装位置和组件等。

（1）双击下载后得到的可执行文件，将显示 PyCharm 的安装界面，如图 1-13 所示。

Follow the instructions of the installation wizard, including agreeing to the license agreement, selecting the installation path and components, etc.

(1)Double-click the downloaded executable file to display the installation interface of PyCharm, as shown in Figure 1-13.

图 1-13　PyCharm 的安装界面

Figure 1-13　Installation Interface of PyCharm

（2）单击 Next 按钮开始安装，选择 PyCharm 的安装位置，建议安装在除 C 盘以外的其他磁盘，如图 1-14 所示。

(2)Click the Next button to start the installation. Select the PyCharm installation location, it is recommended to install it on a disk other than the C drive, as shown in Figure 1-14.

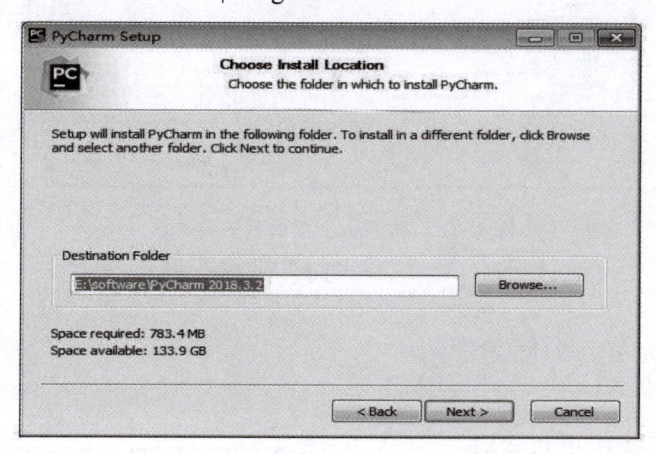

图 1-14　PyCharm 的安装位置界面

Figure 1-14　Installation Location Interface of PyCharm

（3）单击 Next 按钮，进入安装选项界面，在此可以进行安装选项设置，根据自己的需要进行选择，

(3)Click the Next button to display the installation options interface. One can select the installation options according to the needs, it is recommended to select all of

17

建议全部选中，如图1-15所示。 them, as shown in Figure 1-15.

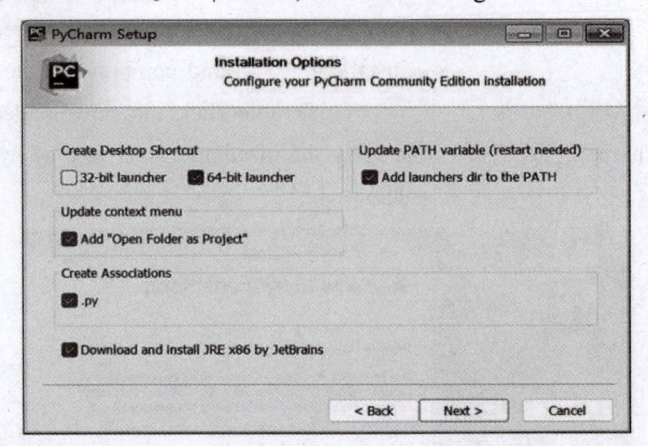

图1-15 PyCharm 的安装选项界面
Figure 1-15　The Installation Options Interface of PyCharm

（4）单击 Next 按钮，进入 PyCharm 的最后安装界面，如图1-16所示，默认安装即可，直接单击 Install 按钮。

(4)Click Next button to show the final installation interface of PyCharm, shown in Figure 1 – 16. Click Install button by default.

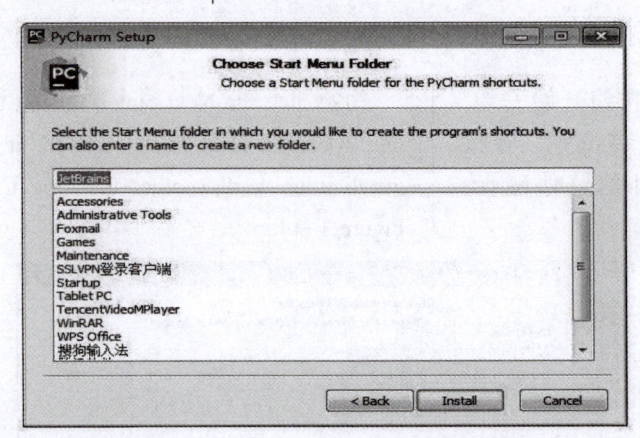

图1-16 PyCharm 的最后安装界面
Figure 1-16　Final Installation Interface of PyCharm

1.4.3 PyCharm 的简单使用
1.4.3　Simple Use of PyCharm

使用 PyCharm 新建项目文件并运行的步骤如下。

The steps for creating a new project file and running it using PyCharm are as follows.

（1）启动 PyCharm，打开 PyCharm 应用程序。如果是首次启动，则可能会看到一个欢迎屏幕，其提供了一些选项，包括新建项

(1)Start PyCharm, open the PyCharm application. If it starts for the first time, one may see a welcome screen with some options, including creating a new project, opening an existing project, and so on.

第1章　Python基础知识
Chapter 1　Basic Knowledge of Python

目、打开已有项目等。

（2）新建项目，在主菜单中选择 File→New Project 命令。在新建项目对话框中，需要选择一个项目解释器。如果已经安装了 Python，则 PyCharm 通常会自动检测到，项目配置界面如图 1-17 所示。如果没有，则需要手动指定 Python 解释器的位置。填写好项目名称和项目位置后，单击 Create 按钮。

(2)To create a new project, select File→New Project command from the main menu. In the Create Project dialog box, one needs to select a project interpreter. If one has already installed Python, PyCharm will automatically detect it usually, the project configuration interface shown in Figure 1-17. If not, one needs to manually specify the location of the Python interpreter. After filling in the project name and project location, click the Create button.

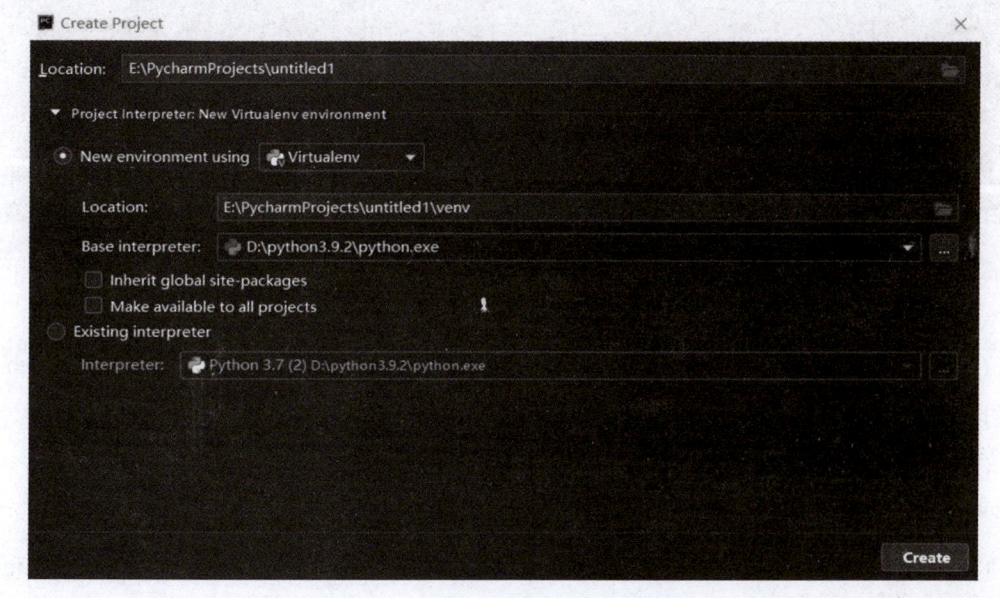

图 1-17　PyCharm 的项目配置界面

Figure 1-17　Project Configuration Interface of PyCharm

（3）新建文件。项目创建完成后，用户会看到一个空的项目结构。要新建文件，需右击项目文件夹，选择 New→Python File 命令（新建 Python 文件），如图 1-18 所示。然后输入文件名（例如，main.py），按〈Enter〉键。

(3)Create a new file. After the project is created, the user will see an empty project structure. To create a new file, one need to right-click on the project folder and select New→Python File command, as shown in Figure 1-18, to create a new Python file. Then enter the file name(for example, main.py) and press〈Enter〉.

（4）编写代码。在新建的 Python 文件中，可以开始编写代码。例如，可以输入以下代码来测试环境。

(4)Writing code. In the newly created Python file, one can start writing code. For example, one can enter the following code to test the environment.

```
print("Hello, PyCharm!")
```

19

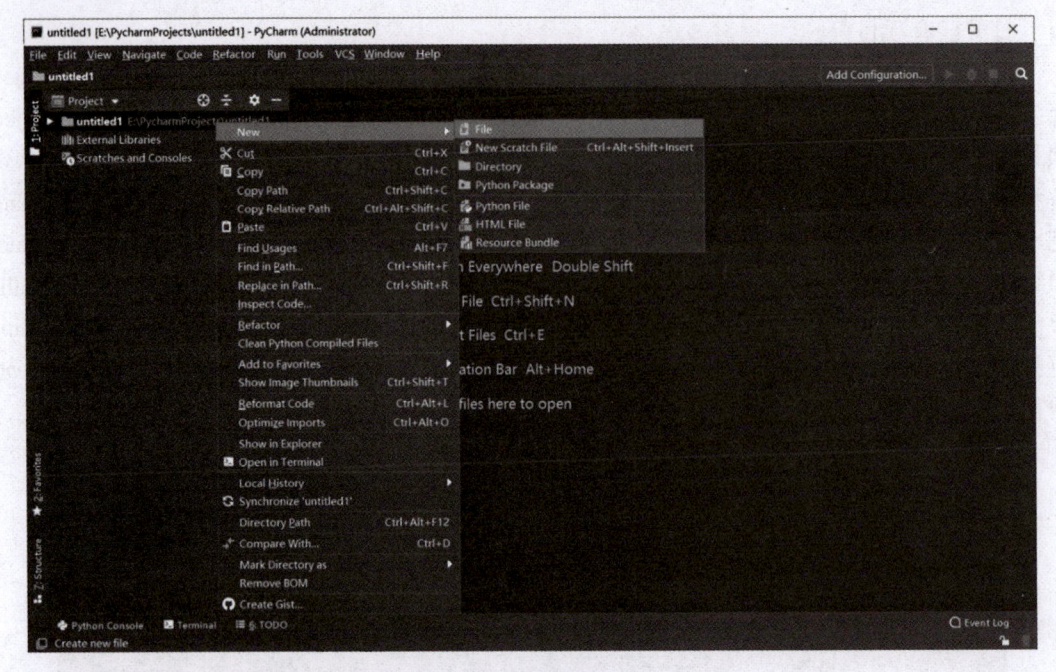

图 1-18 新建项目界面

Figure 1-18 Interface to Create a New Project

(5)运行代码。有以下几种方法可以运行代码。

①右击编辑器中的代码,然后选择"Run 文件名"命令。

②在主菜单中选择 Run→"Run 文件名"命令。

③使用快捷键(通常是〈Shift+F10〉,但这取决于 PyCharm 设置和操作系统)。首次运行代码时,PyCharm 可能会要求用户调试配置。在大多数情况下,默认设置就可以。

(6)查看运行结果。代码运行后,可从 PyCharm 底部的 Run 工具窗口中看到输出结果,如图 1-19 所示。

(5)Running the code. There are several ways to run the code.

①Right-click on the code in the editor and select "Run file name" command.

②Select Run→"Run file name" command in the main menu.

③Use the shortcut key(normally 〈Shift+F10〉, but this depends on the PyCharm settings and operating system). When running the code for the first time, PyCharm may ask the user to debug the configuration. In most cases, the default settings are fine.

(6)After running the code. one can view the output results from the Run tool window at the bottom of PyCharm, as shown in Figure 1-19.

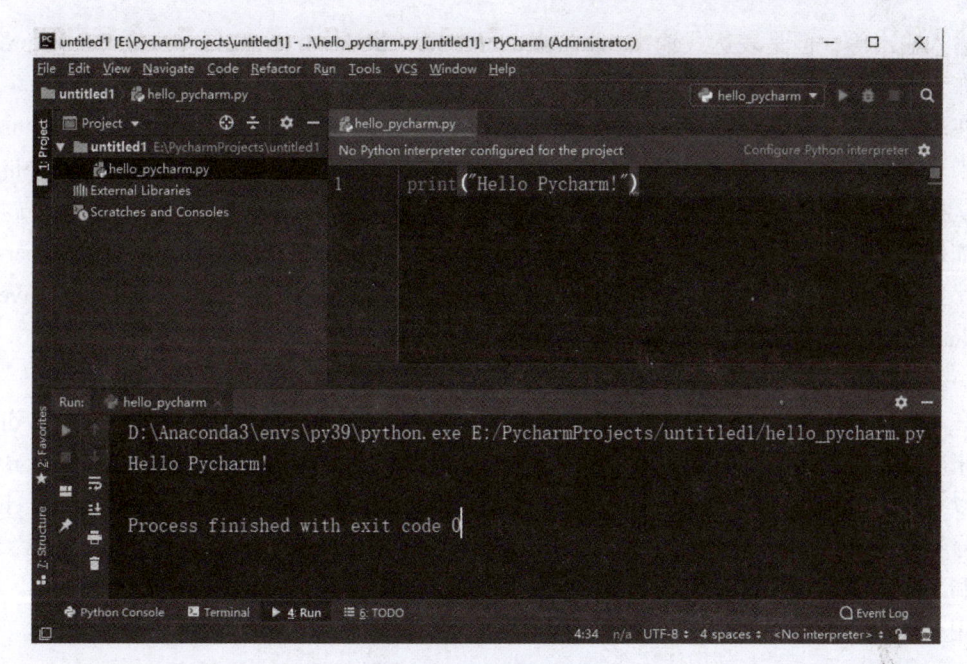

图 1-19　运行成功界面

Figure 1-19　The Interface When Run a Program Successfully

1.5　本章小结
1.5　Summary of This Chapter

程序是一系列为达到特定目的或解决特定任务而按照特定顺序组织的计算机指令。这些指令用各种计算机语言编写，如 Python，并保存在以特定扩展名（如 .py）结尾的文件中。

程序设计语言，从机器语言到汇编语言，再到如今的高级语言，如 Python，它们更接近自然语言，使得程序员能够更高效地编写复杂程序。

Python 开发环境有 IDLE 和 PyCharm 等，在使用 IDLE 编辑器时，注意区分交互式编辑界面和文件式编辑界面的语法区别。能在文件式编辑界面使用的语法，在交互式编辑界面均可以实现。反之，在交互式下可运行测试简单的表达式计

A program is a series of computer instructions organized in a specific order to achieve a specific purpose or solve a specific task. These instructions are written in various computer languages, such as Python, and stored in files ending with a specific extension(such as .py).

Programming languages, from machine language to assembly language to today's advanced languages such as Python, are closer to natural language, allowing programmers to write complex programs more efficiently.

Python development environments include IDLE, PyCharm, etc. When using the IDLE editor, pay attention to the syntax difference between the interactive editing interface and the file editing interface. The syntax that can be used in the file editing interface can be implemented in the interactive editing interface. Conversely, simple expression calculations, input and output, etc. can be

算、输入输出等，但是表达式输出只能在交互式下实现。

在 Python 中，通常使用 input() 函数来获取用户的输入。input() 函数默认接收的数据类型为字符串类型。如果我们需要其他类型的数据计算，比如整数或浮点数，就需要使用 int() 或 float() 等数值类型转换函数。

在 Python 中，可以使用 print() 函数来输出结果，print() 函数的参数可以有 0 个或多个，可以设置分隔符号和结束符号，分别使用 sep 参数和 end 参数来实现，这两个参数使用率较高。

tested and run in the interactive mode, but expression output can only be implemented in this mode.

In Python, the input() function is commonly used to obtain user input. The input() function accepts strings as the default data type. If we need to calculate other types of data, such as integers or floating point numbers, we need to use int() or float() numerical type conversion functions.

In Python, one can use the print() function to display the output. The print() function can have zero or more parameters, and one can set the separator and end characters using the sep and end parameters, respectively. These two parameters are commonly used.

1.6 本章练习
1.6 Exercise for This Chapter

1.1 试着在计算机中安装 IDLE 编辑器，并输出"Hello World"字符串。注意：根据计算机操作系统来进行下载。

1.2 请输出 8 * 3+6 表达式的值，分别在交互式编辑器和文件式编辑器中，试着输出结果。

1.3 使用 print() 函数中的 sep 和 end 参数，输出字符串"1，2，3，4！"。

1.4 使用 input() 函数，输入你的姓名，并输出字符串"Hello 你的姓名！"。

1.1 Try to install the IDLE editor on your computer and output the "Hello World" string. Note: Download according to your computer's operating system.

1.2 Please output the value of the expression 8*3+6, and try to output the result in both the interactive editor and the file editor.

1.3 Use the sep and end parameters in the print() function to output the string "1,2,3,4!".

1.4 Use the input() function to enter your name and output the string "Hello your name!".

本章练习题参考答案

第 2 章　Python 语法基础
Chapter 2　Basic Syntax of Python

2.1　关键字和标识符
2.1　Keywords and Identifiers

2.1.1　关键字
2.1.1　Keywords

关键字是 Python 预先保留的标识符，也称为保留字，如 if、for 等。每个关键字都有特殊的含义，用于控制流程、定义函数和类、进行异常处理等。例如，def 用于定义函数，class 用于定义类，if 和 else 用于条件语句，for 和 while 用于循环等。关键字不能用作变量名、函数名或其他标识符，如果尝试使用它们作为变量名或函数名，则 Python 解释器会抛出一个语法错误。表 2-1 给出了 Python 中的关键字，一共 35 个。

Keywords are Python's reserved identifiers, also known as reserved words, such as if, for, etc. Each keyword has a specific meaning and is used to control flow, define functions and classes, handle exceptions, etc. For example, def is used to define functions, class is used to define classes, if and else are used for conditional statements, for and while are used for loops, and so on. Keywords cannot be used as variable names, function names, or other identifiers. If one attempts to use them as variable names or function names, the Python interpreter will raise a syntax error. Table 2-1 provides the 35 keywords in Python.

表 2-1　Python 中的关键字
Table 2-1　Keywords in Python

False	await	else	import	pass
None	break	except	in	raise
True	class	finally	is	return
and	continue	for	lambda	try

续表

as	def	from	nonlocal	while
assert	del	global	not	with
async	elif	if	or	yield

2.1.2 标识符
2.1.2 Identifiers

Python 中的类名、方法名、对象名、函数名和变量名等，统称为标识符。一个有效的 Python 标识符需要遵守一些基本规则。

（1）由字母、数字和下划线构成：标识符不能以数字开头，但其余部分可以包含数字。例如，my_variable 是一个有效的标识符，而 2myvar 不是。

（2）区分大小写：Python 是大小写敏感的，myVariable 和 myvariable 被视为两个不同的标识符。

（3）不能使用 Python 关键字：表 2-1 中列出的 Python 关键字（如 if、for 等）不能用作标识符。

（4）不能包含特殊字符：除字母、数字和下画线之外，其他字符（如 @、#、$ 等）都不能用于标识符。

（5）尽量使用有意义的名称：虽然 Python 允许使用简短或奇特的标识符，但使用描述性强、易于理解的名称通常会使代码更易于阅读和维护。

下面是一些有效的 Python 标识符示例。

The names of classes, methods, objects, functions, and variables in Python are collectively referred to as identifiers. A valid identifier in Python needs to adhere to some basic rules.

(1) Consist of letters, numbers, and underscores: An identifier cannot start with a number, but the rest can contain numbers. For example, my_variable is a valid identifier, while 2myvar is not.

(2) Case sensitive: Python is case-sensitive, myVariable and myvariable are considered as two different identifiers.

(3) Cannot use Python keywords: Python keywords listed above(such as if, for, etc.) cannot be used as identifiers.

(4) Cannot contain special characters: Apart from letters, numbers, and underscores, other characters(such as @, #, $, etc.) cannot be used in identifiers.

(5) Use meaningful names where possible: Although Python allows the use of short or quirky identifiers, using descriptive and easily understandable names typically makes the code more readable and maintainable.

Here are some examples of valid Python identifiers.

```
my_variable=10
MyClass=type('MyClass')
MY_CONSTANT=52
```

下面是一些无效的 Python 标识符示例。

Here are some examples of invalid Python identifiers.

```
123myvar      #start with a number
if            #use a Python keyword
my-variable   #use a hyphen not underscore
$ my_var      #use other symbols
```

2.2 对象和变量
2.2 Objects and Variables

2.2.1 对象
2.2.1 Objects

在 Python 中，对象是一个核心概念，它是数据和功能的组合体。

Python 是一种面向对象的编程语言，这意味着它的设计是围绕着对象进行的。Python 中的对象可以看作是一个具有属性和方法的实体，这些属性和方法共同描述了对象的状态和行为。对象的基本属性有类型、身份标识和值 3 个。

（1）类型：每个对象都属于一个特定的类型，类型定义了对象可以拥有的属性和方法。例如，整数对象属于 int 类型，字符串对象属于 str 类型。在 Python 中可以使用 type()函数来获取一个对象的类型。

（2）身份标识：对象在内存中的唯一地址。在 Python 中，可以使用 id()函数来获取一个对象的身份标识。

（3）值：对象存储的数据。例如，一个整数对象的值可能是 50，一个字符串对象的值可能是 "Hello, world!"。

In Python, an object is a core concept that combines data and functionality.

Python is an object-oriented programming language, which means its design is based on objects. In Python, objects can be seen as entities with attributes and methods, where these attributes and methods collectively describe the object's state and behavior. The fundamental three attributes of an object are its type, identity, and value.

(1) Type: Each object belongs to a specific type, which defines the attributes and methods that the object can have. For example, integer objects belong to the int type, and string objects belong to the str type. In Python, one can use the type() function to get the type of an object.

(2) Identity: Identity is the unique address of an object in memory. In Python, one can use the id() function to get the identity of an object.

(3) Value: Value is the data stored in an object. For example, the value of an integer object may be 50, and the value of a string object may be "Hello, world!".

2.2.2 变量
2.2.2 Variables

在 Python 中，变量是一种用于存储数据值的标识符。与许多其他编程语言不同，Python 中的变量不需要预先声明类型，它们会自动根据赋给它们的值来获取类型。Python 是一种动态类型语言，这意味着变量的类型可以在程序运行时改变。

为了在实际编程中更方便地引用和操作对象，Python 程序员经常会给对象附加一个标签，这个标签称为名字（name）。这个名字的作用与其他编程语言中的变量相似，它为用户提供了一个引用对象的标识符。在 Python 中，为了保持用法的连贯性和易于理解，习惯上将这种名字称为变量。

In Python, variables are identifiers used to store data values. Unlike many other programming languages, variables in Python do not need to be declared with a specific type, they will automatically take on the type of the value assigned to them. Python is a dynamically typed language, which means that the type of variables can change during program execution.

To conveniently reference and manipulate objects in actual programming, Python programmers often assign a label to objects, which we call a name. This name serves a similar purpose to variables in other programming languages, providing us with an identifier to reference the object. In Python, for consistency and ease of understanding, we commonly refer to these labels as variables.

2.2.3 赋值
2.2.3 Assignment

Python 中的对象具有类型、身份标识和值这 3 个固有属性，而变量则是对这些对象进行标识和引用的名字。这种设计使得 Python 编程更加灵活和直观。变量和对象绑定是通过赋值符号（=）操作完成的，赋值操作的格式如下：

Objects in Python have three inherent attributes: type, identity and value, while variables are names that identify and refer to these objects. This design makes Python programming more flexible and intuitive. The binding of variables to objects is done through the assignment operator(=), and the format of an assignment operation is as follows:

```
Variable name=object
```

例如：

For example：

```
a=10
pi=3.14
```

将变量 a 和对象 10 进行绑定，变量 pi 和对象 3.14 进行绑定，通过 a 和 pi 可以访问绑定对

Bind variable a to object 10, bind variable pi to object 3.14, through a and pi, one can access the type, identity and value of the bound objects.

象的类型、身份标识和值。

```
print(type(a))
print(id(a))
print(a)
print(type(pi))
print(id(pi))
print(pi)
```

输出：　　　　　　　　　　　Output:

```
<class 'int'>
140730815793880
10
<class 'float'>
2892505144560
3.14
```

需要明确的是，在 Python 中，变量仅仅是一个名字，通过赋值符号(=)将变量和对象关联起来。而变量和对象仅是绑定关系，变量不存储对象。

It should be clear that in Python, a variable is just a name that is linked to an object through the assignment operator (=). The relationship between variables and objects is purely binding, and variables do not store objects.

2.2.4 多变量同步赋值
2.2.4 Multiple Variables Simultaneous Assignment

在 Python 中，还可以使用多变量同步赋值来一次性给多个变量赋值。这种赋值方式通常使用元组或列表来实现，也可以通过字符串和 range 对象来完成。要求等号右侧元素的数量要和等号左侧变量的数量保持一致。

In Python, it is possible to use multiple variables assignment to assign values to multiple variables at once. This type of assignment is typically implemented using tuples or lists, and it can also be achieved using strings and range objects. It is important to ensure that the number of elements on the right side of the equal sign matches the number of variables on the left side of the equal sign.

```
a,b,c=(1,2,3)
print(a,b,c,sep=',')
a,b,c=[4,5,6]
print(a,b,c,sep=',')
a,b='你好'
print(a,b,sep=',')
a,b,c,d=range(2,10,2)
print(a,b,c,d,sep=',')
```

输出：

```
1,2,3
4,5,6
你,好
2,4,6,8
```

Output:

2.2.5 变量命名规范
2.2.5 Variable Naming Conventions

因为变量仅仅是一个名字（标识符），所以很多初学者认为使用 a 或 b 这样的简单字母命名变量可以方便地编写出计算机可以正确执行的程序。虽然这样使用在语法上没有问题，但有经验的程序员并不会使用这样一些无意义的字母作为变量名。一般来说，他们会为每个对象起一个简洁且能清晰表达对象意义的名字，以使自己编写的程序可以让其他程序员花尽可能少的时间便能阅读和理解。

Python 变量命名规范遵循前面介绍的标识符命名规范，此外，还需要注意以下要求。

（1）小写字母和下画线：变量名应该全部小写，单词之间用下画线分隔。这种命名方式被称为小写加下画线。例如，this_is_a_long_variable_name。

（2）常量使用大写字母和下划线：常量（即在程序运行期间值不会改变的变量）应该使用大写字母和下划线命名。例如，MAX_VALUE、PI。

（3）不要使用 Python 内置函数或类的名称：尽管技术上可以覆盖内置函数或类的名称，但这样做是不推荐的，因为它可能导致代码难

Since a variable is just a name(identifier), many beginners believe that using simple letters like a or b to name variables can help write programs that computers can execute correctly. Although there is no problem with this usage in terms of syntax, experienced programmers do not use such meaningless letters as variable names. Generally, they give each object a concise and clear name that expresses the meaning of the object, so that the programs they write can be read and understood by other programmers in as little time as possible.

Python variable naming conventions follow the identifier naming rules mentioned earlier, and it is also important to consider the following requirements.

(1) Lowercase letters and underscores: Variable names should be all lowercase, with words separated by underscores. This naming convention is called lowercase with underscores. For example, this_is_a_long_variable_name.

(2) Constants use uppercase letters and underscores: Constants(variables whose values will not change during program execution) should be named using uppercase letters and underscores. For example, MAX_VALUE, PI.

(3)Do not use the names of Python built-in functions or classes: Although it is technically possible to override the names of built-in functions or classes, it is not recommended as it can make the code difficult to understand

以理解和维护。

(4) 保持一致性：在项目的整个代码库中，应保持一致的命名风格。这有助于提高代码的可读性和可维护性。

遵循这些规范可以使 Python 代码更易于阅读和理解，同时也符合 Python 社区广泛接受的编码习惯。请注意，虽然 Python 解释器不会强制要求遵循这些命名规范，但遵循它们是一种良好的编程实践。

and maintain.

(4) Maintain consistency: Throughout the codebase of a project, it is important to maintain a consistent naming style. This helps improve the readability and maintainability of the code.

Following these conventions can make Python code easier to read and understand, while also aligning with widely accepted coding practices in the Python community. Please note that although the Python interpreter does not enforce these naming conventions, following them is considered good programming practice.

2.3 数值类型
2.3 Numerical Types

在 Python 3 中可参与数学运算的数值类型主要有整数(整型)、浮点数(实数)和复数类型，它们的关键字分别是 int、float 和 complex，它们也被称为基本数据类型。

The main numerical types that can participate in mathematical operations in Python 3 are integers(int), floating-point numbers(float) and complex numbers. Their keywords are int, float and complex respectively, and they are also referred to as basic data types.

2.3.1 整数类型
2.3.1 Integer Type

在 Python 中，整数是基本的数值数据类型之一，是不包含小数点的数字，包括十进制的 0、正整数和负整数以及其他进制的整数。例如，123、-456、0b1001（二进制）、0o172（八进制）、0x12af（十六进制），用于表示没有小数部分的数。Python 的整数类型没有大小限制，因此它们可以变得非常大且不会导致溢出，这是 Python 相对于其他编程语言的一个优势。Python 中整数有 4 种表示形式，如表 2-2 所示。

In Python, integers are one of the fundamental numerical data types that do not contain decimal points, including decimal 0, positive integers, negative integers and integers in other number systems. For example, 123, -456, 0b1001 (binary), 0o172 (octal), 0x12af (hexadecimal). Integers are used to represent numbers without a decimal part. Python's integer type has no size limit, they can become very large without causing overflow, which is an advantage of Python compared to other programming languages. There are 4 ways to represent integers in Python as shown in Table 2-2.

表2-2 Python 中整数的不同表示方式

Table 2-2　Different Representations of Integers in Python

进制种类 Types of Numeral Systems	引导符号 Prefix Marks	描述与示例	Description and Examples
十进制 decimal	none	由字符 0 到 9 组成，遇 10 进 1，如 99、156	Composed of characters 0 to 9, with each 10 rolling over to 1, such as 99, 156
二进制 binary	0b or 0B	由字符 0 和 1 组成，遇 2 进 1，如 0b1010、0B1111	Composed of characters 0 and 1, with each 2 rolling over to 1, such as 0b1010, 0B1111
八进制 octal	0o or 0O	由字符 0 到 7 组成，遇 8 进 1，如 0o107、0O777	Composed of characters 0 to 7, with each 8 rolling over to 1, such as 0o107, 0O777
十六进制 hexadecimal	0x or 0X	由字符 0 到 9 及 a、b、c、d、e、f 或 A、B、C、D、E、F 组成，遇 16 进 1，如 0xFF、0X10A	Composed of characters 0 to 9 as well as a, b, c, d, e, f, or A, B, C, D, E, F, with each 16 rolling over to 1, such as 0xFF, 0X10A

```
a=0b1011
b=0o127
c=1234
d=0x12ff
print(a,b,c,d,sep=',')
```

输出： | Output:

11,87,1234,4863

在 Python 3 中，整数的范围几乎是无限的，这是由于它采用了动态的整数表示。这意味着只要计算机的内存足够 Python 能够处理非常大的整数。因此，Python 能够存储和计算远超其他语言能够处理的整数大小。

Factorial(n) 是标准库 math 库中计算阶乘的函数，利用它可以计算正整数 n 的阶乘。在其他语言中，较大数字的阶乘计算是一个复杂的任务。而在 Python 中，可以用以下 2 行代码完成任意非负整数的阶乘运算，而且所得到的结果是完全准确的。

In Python 3, the range of integers is almost infinite due to its adoption of dynamically sized integer representation. This means that Python can handle very large integers as long as the computer's memory is sufficient. Therefore, Python can store and compute integer sizes far beyond what other languages can handle.

Factorial(n) is a function in the standard library math module that calculates the factorial. Using it, one can compute the factorial of a positive integer n. In other languages, computing the factorial of large numbers is a complex task. While in Python, it can be done with just 2 lines of code for any arbitrarily non-negative integer, and the result obtained is completely accurate.

第2章　Python语法基础
Chapter 2　Basic Syntax of Python

```
import math                    #Importing the math library
print(math.factorial(100))     #Calculating the factorial of 100 with the math library
```

输出：　　　　　　　　　　　　　　　　Output：

```
93326215443944152681699238856266700490715968264381621468592963895217599993229915608941463976156518286253697920827223758251185210916864000000000000000000000000
```

2.3.2　布尔类型
2.3.2　Boolean Type

在 Python 中，布尔值（逻辑值）属于整数的子类型，类型关键字为 bool。布尔值有 True 和 False 两个常量对象。它们被用来表示逻辑上的"真值"或"假值"。比较运算的结果是逻辑值，逻辑运算的操作数在运算时表现为逻辑值。在参与数值运算时，布尔值分别以整数"0"和"1"为值参与运算。

In Python, Boolean values(logical values) belong to the subtype of integers, with the type keyword being bool. There are two constant objects, True and False, used to represent "true" and "false" logically. The results of comparison operations are logical values, and the operands in logical operations behave as logical values. When participating in numerical operations, Boolean values are treated as integers with values of "0" and "1".

```
print(False+10)        #Output 10
print(True+10)         #Output 11
```

bool()函数是一个内置函数，用于将对象的值转换为布尔值。除以下这 3 类对象外，其他对象的逻辑值均是 True。3 类对象的布尔值是 False。

(1)被定义为假值的常量，False 和 None。

(2)任何数值型的数字 0，如 0、0.0、complex(0)。

(3)空的序列和多项集：空字符串''、空列表 []、空元组 ()、range(0)、空字典 {}和空集合 set()。

The bool() function is a built-in function in Python used to convert the value of an object into a Boolean value. Apart from the following three categories, the logical value of other objects is considered True. Here are the three categories of objects with a boolean value of False.

(1)Constants explicitly defined as False and None.

(2)Any numeric zero value, such as 0, 0.0, complex(0).

(3)Empty sequences and collections: empty string '', empty list [], empty tuple(),range(0), empty dictionary {} and empty set().

```
print(bool(None))
print(bool(0))
print(bool(''))
print(bool([]))
print(bool(3))
```

```
print(bool('abcd'))
print(bool([1,2]))
```

输出: | Output:

```
False
False
False
False
True
True
True
```

2.3.3 浮点数类型
2.3.3 Floating-point Type

在 Python 中，浮点数有两种表示方式：十进制表示方式和科学计数法表示方式。

(1)十进制表示的浮点数由整数部分、小数点与小数部分组成。小数部分可以没有数字，但必须要有小数点，此时相当于小数部分为0。当其没有小数部分且没有小数点时就变成了整数。例如：3.14、12.、45.0。

(2)科学计数法表示为<a>e<n>的形式，等价于数学中的 $a×10^n$。这里要求 a 和 n 都必须存在，并且 n 必须是整数，e 也可以是大写的 E。例如：2.7e2、2E-3、3.14e5 分别等价于 $2.7×10^2$、$2×10^{-3}$、$3.14×10^5$。

In Python, floating-point numbers have two representation methods: decimal representation and scientific notation representation.

(1)Decimal representation of a floating-point number consists of an integer part, a decimal point, and a decimal part. The decimal part can be empty, but there must be a decimal point, which is equivalent to having a decimal part of 0. When there is no decimal part and no decimal point, it becomes an integer. For example: 3.14, 12., 45.0.

(2)Scientific notation representation is in the form of <a>e<n>, equivalent to $a×10^n$ in mathematics. It requires that both a and n must exist, and n must be an integer. "e" can also be in uppercase "E". For example: 2.7e2, 2E-3, 3.14e5 are equivalent to $2.7×10^2$, $2×10^{-3}$, $3.14×10^5$.

```
a=3.14
b=314e-2
print(a,b,sep=',')
```

输出: | Output:

```
3.14,3.14
```

由于计算机数字的表示采用的是二进制,十进制与二进制转换过程中可能会产生误差,也就是二进制无法精确表示大多浮点数,所以,浮点数是无法保证完全精确的。在 Python 中,浮点数占 8 个字节(64 位)存储空间,能表示的数字范围为 $1.7×10^{-308}$ ~ $1.7×10^{308}$,超过这个范围时会触发溢出异常(OverflowError)。

使用字符串格式化 str.format() 方法时,可以使用":.mf"的方式设置浮点数的小数位数为 m 位,但实际上超过 17 位有效数字后面的数字并不能精确表示。

Due to the fact that computer numbers are represented in binary, during the conversion between decimal and binary, errors may occur, meaning that most of floating-point numbers cannot be accurately represented in binary. Therefore, floating-point numbers cannot guarantee complete accuracy. In Python, floating-point numbers occupy 8 bytes(64 bits) of storage space, and the range of numbers that can be represented is from $1.7×10^{-308}$ to $1.7×10^{308}$. When exceeding this range, an overflow exception(OverflowError) will be triggered.

When using the string formatting methodstr.format(), one can use the ":.mf" method to set the number of decimal places for floating-point numbers to m digits. However, in reality, numbers beyond 17 significant digits cannot be accurately represented.

```
print(pow(809.0,106))        #809.0e106 has a value of 1.748 007 496 839 708 e+308
print(pow(810.0,106))        #OverflowError:(34,'Result too large')
```

```
print('{:.20f}'.format(314159.26535897932384626433827950288419))
print('{:.20f}'.format(3.14159265358979323846264338327950288419))
```

输出:

314159.26535897934809327125
3.14159265358979311599

Output:

314159.26535897934809327125
3.14159265358979311599

2.3.4 复数
2.3.4 Complex Number

Python 支持复数类型和运算,复数由实数部分和虚数部分构成,表示复数可以用 a+bj(j 可以为大写 J)或者使用函数 complex(a,b),其中,复数的实部 a 和虚部 b 都是浮点数。用 real 和 imag 分别获取复数的实部和虚部,用内置函数 abs(a+bj)获得复数的模。

Python supports complex number types and operations. A complex number consists of a real part and an imaginary part. A complex number can be represented as a+bj(where j can be uppercase J) or using the complex(a,b) function, where the real part a and the imaginary part b are both floating-point numbers. One can use real and imag to respectively get the real and imaginary parts of a complex number, and use the built-in function abs(a+bj) to obtain the modulus of the complex number.

```
num=3+4j
print(num.real)             #The real part is 3.0
```

```
print(num.imag)          #The imaginary part is 4.0
print(abs(num))          #The modulus of the complex number is 5.0
```

2.4 字符串类型
2.4 String Type

字符串是 Python 中最常用的数据类型，用于表示文本信息。属于不可变数据类型，意味着不能修改一个字符串中的某个字符，但是可以创建新的字符串。在 Python 中 input()函数获取到的数据、文本文件获得的数据都是字符串数据类型。

在 Python 中表示字符串使用一对单引号(' ')、一对双引号(" ")或一对三引号(''' '''或""" """)为界定符，用引号包围起来的 0 个或多个字符序列就称为一个字符串。包含字符的个数称为字符串的长度，当包含 0 个字符时，称为空字符串。

Strings are the most commonly used data type in Python, used to represent text information. String is an immutable data type, which means that one cannot modify a specific character within a string, but one can create a new string. Data obtained from the input() function or text files in Python are all of the string data type.

In Python, strings are represented using a pair of single quotes(''), a pair of double quotes(" "), or a pair of triple quotes(''' ''' or """ """). Any sequence of zero or more characters enclosed in quotes is called a string. The number of characters in a string is referred to as the length of the string, and when it contains zero characters, it is called an empty string.

2.4.1 字符串创建
2.4.1 String Creation

字符串可以用单引号(')、双引号(")或三引号('''或""")来创建。使用 len()函数可以获取字符串的长度。

Strings can be created using single quotes('), double quotes("), or triple quotes(''' or """). The length of a string can be obtained using the len() function.

```
str0=''
str1='This is a string, which can contain "double quotes"'
str2="This is a string, which can contain 'single quotes'"
#Length is 0, empty string
print(len(''))
#Length is 1, string with a space, 1 space equals 1 character
print(len(' '))
#Length is 16, one space and several punctuation marks
print(len('He said,"hello".'))
#Length is 9, when the string contains single quotes, use double quotes outside
print(len("I'm lilei"))
```

```
#Length is 14
print(len('I love Python!'))
```

用三引号引起来的字符也可以作为字符串来进行处理，字符串内容中可以包含单引号、双引号和回车符，三引号引起来的字符串在输出时可以保持原有格式输出。下述程序可以把一首宋词以字符串形式赋值给变量 poem，输出时会保留原文中的换行等格式。

Characters enclosed in triple quotes can also be treated as strings, and the string content can include single quotes double quotes and carriage returns. Strings surrounded by triple quotes can maintain their original formatting when output. The following program assigns a poem in the form of a string to the variable "poem", and when output, it will retain the original line breaks and other formatting.

```
#Triple quotes can be used for strings to preserve the original format unchanged
poem='''
Quiet Night Thoughts
Before my bed, the bright moonlight,
I suspect it to be frost on the ground.
I lift my head to gaze at the clear moon,
And lower it to think of my hometown.
'''
print(poem)
```

输出：

Output:

```
Quiet Night Thoughts
Before my bed, the bright moonlight,
I suspect it to be frost on the ground.
I lift my head to gaze at the clear moon,
And lower it to think of my hometown.
```

三引号也用于 Python 的注释，称为多行注释。当三引号单独作为一条语句出现时，需要独占一行，按注释处理。当把三引号引起来的内容赋值给变量或作为函数的参数时，按字符串处理。

Triple quotes are also used for comments in Python, known as multiline comments. When triple quotes appear as a standalone statement, they need to occupy a line by themselves and will be treated as comments. When the content enclosed in triple quotes is assigned to a variable or used as a function parameter, it is treated as a string.

2.4.2 转义字符
2.4.2 Escape Characters

在 Python 中，转义字符用于表示特殊字符或不可打印的字符。它们以反斜杠(\)开头，后面跟着一个或多个字符来表示特定的值。表 2-3

In Python, escape characters are used to represent special characters or non-printable characters. They start with a backslash(\) followed by one or more characters to represent a specific value. Table 2-3 lists some com-

列出了一些常用的转义字符。 | monly used escape characters.

表 2-3 常用的转义字符

Table 2-3 Commonly Used Escape Characters

转义字符 Escape character	描述 Description	示例	Example
\n	Line break	用于行末，表示输出到当前位置结束本行，后面字符在新的一行输出	Used at the end of a line, indicates the end of the current line when output, with the following characters being output on a new line
\r	Carriage return	将光标移动到当前行的开头，而不移动到下一行	Move the cursor to the beginning of the line without moving to the next line
\t	Horizontal tabulation	功能与键盘上〈Tab〉键相同，光标水平移动若干个字符	Function similar to the 〈Tab〉 key on the keyboard, moving the cursor horizontally by a certain number of characters
\(at the end of the line)	Line continuation character	为避免一行太长，排版时在前一行末尾加"\"，将下一行内容接在前行末尾	To avoid lines becoming too long, add '\' at the end of the previous line when formatting, to continue the next line from the end of the previous line
\\	Backslash symbol	用于在字符串中输出一个反斜杠"\"	To output a backslash '\' in a string
\'	Single quote	用于在字符串中输出一个单引号	To output a single quote in a string
\"	Double quote	用于在字符串中输出一个双引号	To output a double quote in a string
\b	Backspace	使光标回退一格，清除前面一个字符	Move the cursor back one space and erase the preceding character

```
print("Hello,\nWorld!")
print('He said, \"Hello, world!\"')
```

输出 | Output:

```
Hello,
World!
He said, "Hello, world!"
```

2.5 数据类型转换
2.5 Data Type Conversion

在程序设计过程中，经常需要 | In the process of programming, it is often necessary

对数据类型进行转换，例如，input() 函数获取的数据类型是字符串类型，获取的数据在参与数值运算时就需要进行转换，保证运算结果的正确性。

（1）int(x)可将整数字符串转换成整数，或截取浮点数的整数部分。

to convert data types. For example, the data obtained from the input() function is in string type, so conversion is needed when these data are involved in numerical operations to ensure the accuracy of the computation results.

(1)The int(x) can convert an integer string into an integer, or truncate the integer part of a floating-point number.

```
#The string '5' is converted to a decimal integer. The output is 5
print(int('5'))
#Convert the floating-point number to an integer by retaining only the integer part. The output is 3
print(int(3.14))
```

实际上，int()函数的完整形式为int(x,base=10)，它不仅可以把浮点数转换成整数，把十进制整数字符串转换成整数，还可以把其他进制字符串形式转换成数值类型。base可以取的值包括0、2~36中的整数，当base取值为0时，系统根据字符串前的进制引导符确定该数的进制，例如：

Actually, the complete form of the int() function is int(x, base = 10), it can not only convert floating-point numbers to integers and convert decimal integer strings to integers, but also convert other base string forms into numerical types. The base can take on values including integers from 0, 2 to 36. When the base value is 0, the system determines the base of the number based on the leading indicator in the string, for example:

```
#The prefix '0o' indicates that the number is in octal representation. The output is 71
print(int('0o107',base=8))
print(int('0o107',base=0))
#The prefix '0x' indicates that the number is in hexadecimal representation. The output is 263
print(int('0x107',base=16))
print(int('0x107',base=0))
#The prefix '0b' indicates that the number is in binary representation. The output is 9
print(int('0b1001',base=2))
print(int('0b1001',base=0))
```

输出：

Output:

```
71
71
263
263
9
9
```

需要注意的是,int()函数只能将整数字符串或浮点数转成整数,而不能将浮点数字符串转成整数,例如,int('3.14'),系统会返回 ValueError 异常。

Please note that the int() function can only convert integer strings or floating point numbers to integers, and cannot convert floating point string to an integer. For example, int('3.14') will raise a ValueError exception.

```
print(int('3.14'))    #ValueError: invalid literal for int() with base 10: '3.14 '
```

(2)float(x)函数可以将整数 x 或浮点数字符串转换成浮点数,也可以将整数字符串转换成浮点数。

(2)The float(x) function can convert an integer x or a floating-point number string to a floating-point number, and can also convert an integer string to a floating-point number.

```
print(float(3))
print(float('4.56'))
print(float('5'))
```

输出:

Output:

```
3.0
4.56
5.0
```

(3)complex(x)函数可以将浮点数 x 转换成一个复数,实数部分是 x,虚数部分是 0,x 还可以是一个可以转换为数字或复数的字符串。complex()函数可以接收两个参数,complex(x[,y])的功能是将 x 和 y 转换为一个复数,实数部分为 x,虚数部分为 y。

(3) The complex (x) function can converts the floating-point number x into a complex number, with the real part being x and the imaginary part being 0, x can also be a string that can be converted to a number or complex number. The complex() function can also take two parameters, complex(x[, y]), where x and y are converted to a complex number, with x as the real part and y as the imaginary part.

```
print(complex(3))
print(complex('3+4j'))
print(complex(3,4))
```

输出:

Output:

```
(3+0j)
(3+4j)
(3+4j)
```

(4)str(x)函数可以将对象 x 转为字符串类型,这是非常有用的,因为字符串是 Python 中处理文本

(4)The str(x) function can convert the object x into a string type, which is very useful because strings are the primary way to handle text in Python, and many opera-

第2章　Python语法基础
Chapter 2　Basic Syntax of Python

的主要方式，而且许多操作都要求数据以字符串形式存在。虽然 str() 可以将对象转换为字符串，但它并不会改变原始对象的类型，它返回的是一个新的字符串对象。

tions require data to be in string form. While str() can convert an object into a string, it does not change the original object's type, it returns a new string object.

```
num=1234
str_num=str(num)
print(str_num,type(str_num),sep=',')
float_num=3.14159
str_float_num=str(float_num)
print(str_float_num,type(str_float_num),sep=',')
```

输出：

Output:

```
1234,<class 'str'>
3.14159,<class 'str'>
```

（5）eval(x)函数可以将数值型的字符串对象 x 转换为其对应的数值。例如，x 为整数字符串"3"时，转换结果为数值 3；当 x 为浮点数字符串"3.0"时，转换结果为数值 3.0。eval(x)函数的功能可以总结为：脱掉字符串两侧的引号，执行剩余的语句。因此，借助 eval(x)函数和 input()函数还可以完成多变量同步赋值操作。

(5) The eval(x) function can converts a numeric string object x into its corresponding numerical value. For example, when x is an integer string "3", the conversion result is the numerical value 3; when x is a floating-point string "3.0", the conversion result is the numerical value 3.0. The functionality of the eval(x) function can be summarized as: Removing the quotes on both sides of the string and executing the remaining statement. Thus, with the help of the eval(x) function and the input() function, multiple variable synchronous assignment operations can also be completed.

```
a=eval('10')
print(a)
b=eval('10.0')
print(b)
c=eval('3+7')
print(c)
#input 3,4
x,y=eval(input())
print(x,y)
```

输出：

Output:

```
10
10.0
```

39

```
10
3 4
```

【例 2-1】计算矩形面积。矩形的面积等于其长与宽的乘积，用户输入长和宽的值，按输入要求编程计算矩形的面积。输入要求如下。

（1）在 Python 中 input()函数获取的数据类型是字符串类型，而字符串无法参与算数运算，所以在程序中需要将输入的字符串转为数值类型。当用户的输入确定是整数时，程序中可以用 int()函数将输入转为整数类型，计算结果也是整数类型。

[Example 2-1] Calculating the area of a rectangle. The area of a rectangle is equal to the product of its length and width. Program the calculation of the rectangle's area based on user input of the length and width values. The input requirements are as follows.

(1)The data type obtained by the input() function in Python is a string type, and strings cannot participate in arithmetic operations. Therefore, it is necessary to convert the input string to a numerical type in the program. When the user's input is confirmed to be an integer, the int() function can be used in the program to convert the input to an integer type, and the calculation result will also be an integer type.

```
#Please input the length and width of the rectangle represented as integers.
#Calculate and output the area of the rectangle
#To convert the input into an integer, for example, enter 5. Use the int() function
width=int(input('Please input the width:'))
#To convert the input into an integer, for example, enter 6. Use the int() function
length=int(input('Please input the length:'))
#To calculate the area using the area formula
area=width*length
#Output 30
print(area)
```

2-1-1

（2）当用户的输入是浮点数时，可以用 float()函数将获取的浮点数字符串转为浮点数类型。当输入为整数时，input()函数获取的是整数字符串，当然也会被转为浮点数，计算结果也是浮点数。

(2)When the user's input is a floating-point number, the obtained floating-point number string can be converted to a floating-point type using the float() function. When the input is an integer, the input() function obtains an integer string, which will also be converted to a floating-point number, and the calculation result will be a floating-point number as well.

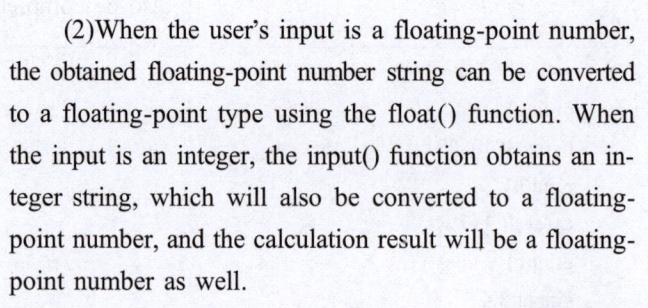

```
#Input the length and width of a rectangle represented by floating-point numbers,
#calculate and output the area of the rectangle
#Convert the input to a floating-point number using float(), input 3.6
width=float(input('Please input the width:'))
#Convert the input to a floating-point number using the float() function, input 6.3
length=float(input('Please input the length:'))
```

2-1-2

```
area=width*length
#Output 22.68
print(area)
print(round(area,2))
```

（3）当不确定用户输入的是整数还是浮点数时，如果想保证计算结果与输入的数据类型一致，可以使用 eval() 函数，该函数在将输入转换为可计算对象时，会保持数据类型与输入一致。

当输入整数时，转换后还是整数；当输入浮点数时，转换后还是浮点数。

(3) When unsure whether the user input is an integer or a float, if one wants to ensure that the calculation result matches the data type of the input, one can use the eval() function. This function, when converting the input into a computable object, will maintain the data type consistent with the input.

If one inputs an integer, it will still be an integer after conversion; if one inputs a float, it will still be a float after conversion.

```
#Input the length and width of a rectangle represented by positive numbers, calculate
#and output the area of the rectangle
#Use the eval() function to convert the input to a numerical type
width=eval(input('Please input the width:'))
#Use the eval() function to convert the input to a numerical type
length=eval(input('Please input the length:'))
#Calculate the area using the area formula
area=width*length
print(area)
```

2-1-3

2.6 运算符及表达式
2.6 Operators and Expressions

在程序设计中，运算符和表达式扮演着至关重要的角色，它们是实现各种计算、比较和逻辑判断的基础。在 Python 3 中，这些运算形式非常丰富和灵活，能够满足各种编程需求。Python 3 支持数值运算、比较（关系）运算、成员运算、布尔（逻辑）运算、身份运算、位运算和真值测试等运算形式。

表达式是由运算符和操作数组合而成的语句，用于计算并返回一个值。它们可以出现在程序

Operators and expressions play a crucial role in program design, as they are the foundation for implementing various calculations, comparisons, and logical judgments. In Python 3, these forms of operations are very rich and flexible, capable of meeting various programming needs. Python 3 supports numerical operations, comparison(relational) operations, membership operations, boolean(logical) operations, identity operations, bitwise operations, truth value testing, and other forms of operations.

An expression is a statement formed by combining operators and operands, used to calculate and return a value. They can appear in different parts of a program,

的不同部分，比如赋值语句的右侧、条件语句中，或者作为函数的参数等。表达式的写法与数学中的表达式稍有不同，需要按照程序设计语言规定的方式构造表达式。

such as the right side of assignment statements, within conditional statements, or as arguments to functions. The syntax of expressions differs slightly from expressions in mathematics and needs to be constructed according to the rules specified by the programming language.

2.6.1 算术运算符和表达式
2.6.1 Arithmetic Operators and Expressions

Python 中包括加(+)、减(-)、乘(*)、除(/)、整除(//)、取余(%)等基本的算术运算，以及指数运算(**)等。

In Python, there are basic arithmetic operations such as addition(+), subtraction(-), multiplication(*), division(/), floor division(//), modulo(%), and exponentiation(**).

这些运算符用于处理数值类型的数据，如整数和浮点数。它们的优先级为：指数运算符最高，其次是乘、除、整除和取余运算符(优先级相同)，最低的是加和减运算符(优先级相同)。具体功能如表2-4所示。

These operators are used to manipulate numerical data types, such as integers and floating-point numbers. Their precedence follows the order from highest to lowest: exponentiation operator has the highest precedence, followed by multiplication, division, floor division, and modulo operators(with equal precedence), and finally addition and subtraction operators(with equal precedence). Their specific functionalities are shown in Table 2-4.

表 2-4 算数运算符(表中 a=8, b=5)
Table 2-4 Numerical Operators(a=8, b=5 in the table)

运算符 Operator	功能描述	Functional Description	示例 Example
+	加：两个数相加	Addition: adding two numbers	print(a+b) #The result is 13
-	减：两个数相减	Subtraction: subtracting two numbers	print(a-b) #The result is 3
*	乘：两个数相乘	Multiplication: multiplying two numbers	print(a*b) #The result is 40
/	除：两个数相除	Division: dividing two numbers	print(a/b) #The result is 1.6
//	整除：返回商的整数部分，向下取整	Floor division: returning the integer part of the quotient, rounding down	print(a//b) #The result is 1 print(-10//4) #The result is -3

第2章　Python语法基础
Chapter 2　Basic Syntax of Python

续表

运算符 Operator	功能描述	Functional Description	示例 Example
%	取模：a%b=a-(a//b)*b	Modulo：a%b = a-(a//b)*b	print(a%b)　#The result is 3 print(-10%3)　#The result is 2
**	幂：返回 x 的 y 次幂	Exponentiation：returning x power y	print(a**b)　#The result is 32768

在 Python 中，加、减、乘与数学上同类运算意义相同。除法有以下两种。

（1）精确除法（/）：无论参与运算的数是整数还是浮点数，是正数还是负数，都直接进行除法运算，运算结果的类型都是浮点数。

（2）整除（//）：采用向下取整的算法得到整数结果。所谓向下取整，是对精确除法结果向负无穷大的方向取整。例如，10//3 的结果是 3。需要注意的是，当参与运算的两个操作数都是整型数字时，结果是整型；当有浮点数参与运算时，结果为浮点型。

In Python, addition, subtraction, multiplication have the same meaning with similar mathematical operations in mathematics. There are two types of division.

(1)Exact division(/): Regardless of whether the numbers involved in the operation are integers or floating-point numbers, positive or negative, exact division directly calculates the result and returns a floating-point number.

(2) Floor division (//): Floor division uses the algorithm of rounding down to obtain an integer result. Rounding down means rounding the exact division result towards negative infinity. For example, the result of 10 // 3 is 3. It's important to note that when both operands in the operation are integers, the result is an integer; when there is a floating-point number involved, the result is a float.

```
print(20/5)
print(-10/4)
print(10//4)
print(-10//4)
print(10.0//4)
```

输出：　　　　　　　　　　　　　Output:

```
4.0
-2.5
2
-3
2.0
```

在 Python 中，% 运算符是取模运算，也称为求余运算。取模运算在数论和程序设计中都有着

In Python, the % operator is the modulo operation, also known as the remainder operation. The modulo operation has wide applications in number theory and pro-

43

广泛的应用，例如奇偶数和素数的判断，取任意长度正整数个位上的数字等。取模运算的结果为表达式 a-(a//b)*b 的值。例如，如果 a=10，b=3，那么 a%b 的值是 1。

Python 中使用两个星号"**"表示幂运算，表达式 a**b 表示的是 a 的 b 次幂。幂运算优先级比取反高，如-5**2 的运算顺序与-(5**2)相同，即先进行幂运算，再取反，最终的值为-25。在复杂表达式中适当加括号是较好的编程习惯，既可以确保运算按自己预定的顺序进行，又可以提高程序的可读性和可维护性。当 b 的值小于 1 时，计算的是根号运算。

gramming, such as determining odd and even numbers, prime number checking, extracting digits from positive integers of any length, etc. The result of the modulo operation is the value of the expression a-(a//b)*b. For example, if a=10 and b=3, then the value of a%b is 1.

In Python, the double asterisk "**" is used to represent exponentiation. The expression a**b denotes a power b. The priority of exponentiation is higher than that of negation. For example, the operation of -5**2 is the same as -(5**2), meaning that exponentiation is done first, then negation, resulting in the final value of -25. It is a good programming practice to use parentheses in complex expressions to ensure that operations are carried out in the intended order and improving the readability and maintainability of the program. When the value of b is less than 1, the calculation represents a square root operation.

```
print(-5**2)
print(-(5**2))
print((-5)**2)
print(5**(1/2))
print(5**(1/3))
```

输出：　　　　　　　　　　　　　　Output:

```
-25
-25
25
2.23606797749979
1.7099759466766968
```

【例 2-2】一元二次方程问题求解。一元二次方程问题可以根据求根公式进行求解。现有一元二次方程：$ax^2+bx+c=0$，a、b、c 的值由用户输入，编程求其根。

将求根公式转换为表达式：

$$x1=(-b+(b**2-4*a*c)**(1/2))/(2*a)$$
$$x2=(-b-(b**2-4*a*c)**(1/2))/(2*a)$$

表达式中的乘号不可以省略，分母中的(2*a)的括号不能省略，否则因乘除的优先级相同，会按先

[Example 2-2] Solving a quadratic equation problem. One can solve a quadratic equation problem using the root formula. Given a quadratic equation: $ax^2+bx+c=0$, where the values of a, b, c are provided by the user, program to find its roots.

Convert the root formula into expression:

$$x1=(-b+(b**2-4*a*c)**(1/2))/(2*a)$$
$$x2=(-b-(b**2-4*a*c)**(1/2))/(2*a)$$

The multiplication symbol in the expression cannot be omitted, and the parentheses around (2*a) in the denominator cannot be omitted either. Otherwise, due to the

第2章　Python语法基础
Chapter 2　Basic Syntax of Python

后顺序进行运算，那么结果就是除2再乘a。如果一定要去掉括号的话，可以将(2*a)中的乘号改为除号(/)以保持数学上的运算顺序。分子里(1/2)的括号不可以省略，因为幂运算优先级高于除法运算，所以没有括号时会先计算1次幂，再除2。为避免这个问题，也可以将(1/2)改写为0.5。

equal priority of multiplication and division, the operations will be performed in sequential order, resulting in dividing by 2 and then multiplying by a. If one must remove the parentheses, one can change the multiplication symbol in (2*a) to a division symbol (/) to maintain the mathematical order. The parentheses around (1/2) in the numerator cannot be omitted because exponentiation has a higher priority than division, and without parentheses, it will calculate the exponentiation first and then divide by 2, resulting in an incorrect order of operations. To avoid this issue, one can rewrite (1/2) as 0.5.

```
a,b,c=eval(input('Input a, b and c:'))
x1=(-b+(b*b-4*a*c)**(1/2))/(2*a)
x2=(-b-(b*b-4*a*c)**(1/2))/(2*a)
#Replace 1/2 with 0.5
#x2=(-b-(b**2-4*a*c)**0.5)/(2*a)
print(x1,x2)
```

2-2

输出(假设输入3,8,5)：　　　Output(assuming input 3, 8, 5):

```
3,8,5
-1.0 -1.6666666666666667
```

2.6.2　复合赋值运算符和表达式
2.6.2　Compound Assignment Operators and Expressions

Python中的复合赋值运算是一种结合了赋值和算术运算或位运算的快捷方式。复合赋值运算符可以在一步中完成变量的更新，使得代码更简洁、更易于阅读。常用的复合赋值运算符包括：+=、-=、*=、/=、//=、%=和**=，它们的等价关系如表2-5所示。

Compound assignment operator in Python is a convenient way to combine assignment with arithmetic or bitwise operations. These operators allow for updating variables in a single step, making the code more concise and easier to read. The commonly used compound assignment operators include: +=, -=, *=, /=, //=, %=, and **=. Their equivalence relationships are shown in Table 2-5.

表2-5　常用复合赋值运算符的等价关系
Table 2-5　Equivalence of Commonly used Compound Assignment Operators

表达式 Expression	等价表达式 Equivalent Expression
a+=b	a=a+b

45

续表

表达式 Expression	等价表达式 Equivalent Expression
a-=b	a=a-b
a*=b	a=a*b
a/=b	a=a/b
a//=b	a=a//b
a%=b	a=a%b
a**=b	a=a**b

```
a=10
b=20
a+=b
print(a)
a*=b+5
print(a)
```

输出： | Output:

```
30
750
```

2.6.3 比较运算符和表达式
2.6.3 Comparison Operators and Expressions

在 Python 中，比较运算（也称为关系运算）用于比较两个或多个值，并返回布尔值（True 或 False）来表示比较的结果。Python 支持多种比较运算符，它们可以帮助我们比较数值、序列等数据类型。

Python 中有 6 种比较运算符，包括 2 种一致性比较符（==、!=）和 4 种次序比较符（<、>、<=、>=）。它们的优先级相同，可以连续使用，例如，x<y<=z 相当于同时满足条件 x<y 和 y<=z。比较运算符如表 2-6 所示。

In Python, comparison operations(also known as relational operations) are used to compare two or more values and return a boolean value(True or False) to indicate the result of the comparison. Python supports various comparison operators that can help us compare numerical values, sequences, and other data types.

In Python, there are 6 comparison operations, including 2 equality comparison operations(==, !=) and 4 relational comparison operations(<,>,<=,>=). They have the same priority and can be used consecutively. For example, x<y<=z is equivalent to satisfying both conditions x<y and y<=z simultaneously. Comparison operators are provided as shown in Table 2-6.

表 2-6 比较运算符
Table 2-6　Comparison Operators

运算符 Operator	描述	Description
==	等于：比较 a、b 两个对象是否相等	Equal：Compare two objects a and b to see if they are equal
!=	不等于：比较 a、b 两个对象是否不相等	Not equal：Compare two objects a and b to see if they are not equal
>	大于：返回 a 是否大于 b	Greater than：Return whether a is greater than b
<	小于：返回 a 是否小于 b	Less than：Return whether a is less than b
>=	大于或等于：返回 a 是否大于或等于 b	Greater than or equal to：Return whether a is greater than or equal to b
<=	小于或等于：返回 a 是否小于或等于 b	Less than or equal to：Return whether a is less than or equal to b

```
a,b=3,5
print('a==b: ',a==b)
print('a !=b: ',a !=b)
print('a > b: ',a > b)
print('a < b: ',a < b)
print('a >=b: ',a >=b)
print('a <=b: ',a <=b)
```

输出：　　　　　　　　　　　　　Output:

```
a==b:   False
a!=b:   True
a>b:    False
a<b:    True
a>=b:   False
a<=b:   True
```

2.6.4　身份运算符和表达式
2.6.4　Identity Operators and Expressions

身份运算符用于比较两个对象在内存中地址是否相同。可用 id() 函数获取对象内存地址，若两个对象的内存地址相同，也就是使用 id() 函数获取的值相等，则为同一个对象，若内存地址不同，则是不同的

The identity operator is used to compare whether the memory addresses of two objects are the same. The id() function can be used to obtain the memory address of an object. If the memory addresses of two objects are the same, meaning the values obtained by using the id() function are equal, then they are the same object; if the me-

47

对象。身份运算符的描述如表 2-7 所示。

mory addresses are different, then they are different objects. The description of the identity operator is shown in Table 2-7.

表 2-7 身份运算符的描述
Table 2-7 Description of the Identity Operator

运算符 Operator	描述	Description	示例 Example
is	判断两个标识符是否引用自一个对象	Check if two identifiers reference the same object	x is y is equivalent to id(x)==id(y). If they refer to the same object, it returns True; otherwise, it returns False
is not	判断两个标识符是否引用自不同对象	Check if two identifiers reference different objects	x is not y is equivalent to id(x)!=id(y). If they do not refer to the same object, it returns True; otherwise, it returns False

```
x=10
y=10
z=11
print(id(x),id(y),id(z))
print(x is y,x is not y)
print(x is z,x is not z)
```

输出： | Output:

```
140730866191064 140730866191064 140730866191096
True False
False True
```

2.6.5 成员运算符和表达式
2.6.5 Membership Operators and Expressions

运算符 in 和 not in 用于成员检测。如果 x 是 s 的成员，则 x in s 值为 True，否则为 False。x not in s 返回 x in s 取反后的值。

所有内置序列、集合类型以及字典都支持此运算。对于字典，in 检测其在字典的键中是否存在，而不是检测其在值中是否存在。

对于字符串类型，当且仅当 x 是 y 的子串时，x in y 为 True。空字符串是任何其他字符串的子串，因

The operators "in" and "not in" are used for membership testing. If x is a member of s, then the value of x in s is True, otherwise it is False. x not in s returns the opposite value of x in s.

This operation is supported by all built-in sequences, set types, and dictionaries. For dictionaries, in checks if the key exists in the dictionary, rather than checking if the value exists.

For string types, x in y is True only if x is a substring of y. The empty string is a substring of any other string, so ""in"Python"will return True. The description of

此""in"Python"将返回 True。成员运算符的描述如表 2-8 所示。

the membership operator is shown in table 2-8.

表 2-8 成员运算符的描述
Table 2-8 Description of the Membership Operator

运算符 Operator	描述	Description
in	如果对象在某一个序列中存在，则返回 True，否则返回 False	If the object exists in a sequence, it returns True; otherwise, it returns False
not in	如果对象在某一个序列中不存在，则返回 True，否则返回 False	If the object does not exist in a sequence, it returns True; otherwise, it returns False

```
print('I'  in ['I','love','Python'])         #True
print('I'  in 'I love Python')               #True
print('L'  not in ['I','love','Python'])     #True
print('I'  not in 'I love Python')           #False
str1='I love Python!'
ls=[1,2,3,4,5]
print('love' in str1,'love' not in str1,sep=',')   #True,False
print(2 in ls,6 in ls)                       #True,False
```

2.6.6 逻辑运算符和表达式
2.6.6 Logical Operators and Expressions

Python 支持逻辑运算（布尔运算），运算符包括 and（与）、or（或）、not（非）。

not 优先级最高，or 优先级最低，逻辑运算符的优先级低于比较运算符，高于赋值运算符。

Python 中有 3 类对象的布尔值是 False：False、None；所有类型的数值零；空序列（空列表、空元组、空字符串和 range(0)）和空多项集（空字典和空集合）。所有其他对象的布尔值是 True。逻辑运算符的表达式与功能描述如表 2-9 所示。

Python supports logical operations(boolean operations) with the operators and, or, and not.

The not operator has the highest precedence, or has the lowest precedence. The precedence of boolean operators is lower than comparison operators and higher than assignment operators.

In Python, there are three types of objects whose Boolean value is False: False, None; all types of numeric zero; and empty sequences(empty list, empty tuple, empty string, and range(0)) and empty mappings (empty dictionary and empty set). The Boolean value of all other objects is True. The expressions and function descriptions of the boolean operators are shown in Table 2-9.

表 2-9 逻辑运算符的表达式与功能描述
Table 2-9 Expressions and Function Descriptions of the Boolean Operators

运算符 Operator	表达式 Expression	X 的布尔值 The Boolean Value of x	Y 的布尔值 The Boolean Value of y	结果（布尔值）Result (Boolean Value)	描述	Description
and	x and y	True	True	y(True)	表达式一边有 False 就会返回 False，当两边都是 True 时返回 True	If one side of the expression is False, it will return False; when both sides are True, it will return True
		True	False	y(False)		
		False	True	x(False)		
		False	False	x(False)		
or	x or y	True	True	x(True)	表达式一边有 True 就会返回 True，当两边都是 False 时返回 False	If one side of the expression is True, it will return True; when both sides are False, it will return False
		True	False	x(True)		
		False	True	y(True)		
		False	False	y(False)		
not	not x	True		False	表达式取反，返回值与原值相反	Expression negation, the return value is the opposite of the original value
		False		True		

例如： For example:

```
print(1 and 2)
print(bool(1 and 2))
print(1 and 0)
print(bool(1 and 0))
print(0 and 2)
print(bool(0 and 2))
print(0 and [])
print(bool(0 and []))
print(1 or 2)
print(bool(1 or 2))
print(1 or 0)
print(bool(1 or 0))
print(0 or 2)
print(bool(0 or 2))
print(0 or [])
print(bool(0 or []))
print(not 1)
print(not 0)
```

输出：

Output:

```
2
True
0
False
0
False
0
False
1
True
1
True
2
True
[]
False
False
True
```

事实上，Python 中的 and 和 or 均为短路运算符，对于 and 运算符，如果第一个操作数为 False，则整个表达式的结果为 False，不会评估第二个操作数；如果第一个操作数为 True，则返回第二个操作数的值作为表达式的结果。对于 or 运算符，如果第一个操作数为 True，则整个表达式的结果为 True，不会评估第二个操作数；如果一个操作数为 False，则返回第二个操作数的值作为整个表达式的结果。这表明当逻辑表达式的值已经能够确定时，Python 解释器会停止进一步计算表达式的剩余部分，从而提高效率。

In fact, both "and" and "or" are short-circuit operators in Python. For the "and" operator, if the first operand is False, the result of the entire expression is False, and the second operand will not be evaluated; if the first operand is True, the value of the second operand is returned as the result of the expression. For the "or" operator, if the first operand is True, the result of the entire expression is True, and the second operand will not be evaluated; if the first operand is False, the value of the second operand is returned as the result of the expression. This indicates that when the value of a logical expression can already be determined, the Python interpreter stops further evaluation of the remaining part of the expression, thereby improving efficiency.

2.6.7 运算符优先级
2.6.7 Operator Precedence

Python 中的运算符优先级决定了当多个运算符出现在同一个表达

The operator precedence in Python determines which operator is evaluated first when multiple operators are

式中时，哪个运算符会首先被计算。这遵循一套固定的规则，类似于数学中的运算顺序。不同运算符的优先级不同，编写程序时要注意各运算符的优先级，程序运行时按运算符优先级从高到低进行运算，优先级相同的运算符一般按从左到右的顺序进行运算，只有幂运算（**）、单目运算（例如 not）、赋值运算符和三目运算符例外，它们从右向左执行。

运算符优先级（从高到低的顺序）如表 2-10 所示。我们需要掌握常用的运算符优先级，算数运算符高于比较运算符，比较运算符高于逻辑运算符，逻辑运算符高于赋值运算符。比较运算符、身份运算符和成员运算符优先级相同。

present in the same expression. This follows a fixed set of rules similar to the order of operations in mathematics. Different operators have different precedence levels, so it is important to be mindful of the precedence while writing programs. During program execution, operators are executed from high precedence to low precedence, with operators of the same precedence generally evaluated from left to right, except for exponentiation(**), unary operators (such as not), assignment operators, and ternary operators, which are executed from right to left.

The operator precedence(listed from high precedence to low precedence) is shown in Table 2-10. It is important to understand the common operator precedence, where arithmetic operators take precedence over comparison operators, comparison operators take precedence over logical operators, and logical operators take precedence over assignment operators. Comparison operators, identity operators, and membership operators have the same precedence level.

表 2-10　运算符优先级（从高到低的顺序）
Table 2-10　Operator Precedence（Listed from High Precedence to Low Precedence）

序号 Serial Number	运算符 Operator	描述	Description
1	()、[]、{}	括号表达式，元组、列表、字典、集合显示	parentheses, tuple, list, dictionary, set
2	x[i]，x[m: n]	索引、切片	index, slice
3	**	幂运算	exponentiation
4	~	按位翻转	bitwise flip
5	+x、-x	正、负	positive and negative
6	*、/、//、%	乘法、除法、整除与取模	multiplication, division, floor division, and modulo
7	+、-	加法与减法	addition and subtraction
8	<<, >>	移位	shift operation
9	&	按位与	bitwise AND
10	^	按位异或	bitwise XOR
11	|	按位或	bitwise OR

续表

序号 Serial Number	运算符 Operator	描述	Description
12	<、<=、>、>=、!=、==	比较运算符	comparison operators
	is、is not	身份运算符	identity operators
	in、not in	成员测试	membership testing
13	not x	逻辑非	logical NOT
14	and	逻辑与	logical AND
15	or	逻辑或	logical OR
16	if-else	条件表达式	conditional expression
17	lambda	lambda 表达式	lambda expression
18	:=	海象运算符	walrus operator

```
x=4*2**3
y=(4*2)**3
print(x,y,sep=',')
```

输出: | Output:

32,512

2.7 常用数学函数
2.7 Common Mathematical Functions

Python 内置了一系列与数学运算相关的函数，可以直接使用这些函数。这些函数在安装好 Python 环境后就默认提供，类似于安装完操作系统后，操作系统默认提供了一些诸如计算器、浏览器等程序，方便用户使用。

下面给出几个常用函数的描述与示例。

(1) abs(x) 返回 x 的绝对值，x 可以是整数或浮点数。当 x 是复数时返回复数的模。

Python has built-in a series of math-related functions that can be used directly. These functions are provided by default once Python is installed, similar to how an operating system provides programs like a calculator and a browser after installation for users' convenience.

Below are descriptions and examples of several commonly used functions.

(1)abs(x) returns the absolute value of x, x can be an integer or a floating-point number. When x is a complex number, it returns the modulus of the complex number.

53

```
print(abs(-5))              #Return the absolute value of an integer, output 5
print(abs(-3.14))           #Return the absolute value of a floating-point number, output 3.14
```

（2）divmod(a,b) 以元组形式返回整数商和余数，相当于(a//b, a%b)。

(2)divmod(a, b) returns the integer quotient and remainder in the form of a tuple, equivalent to (a//b, a%b).

```
#Return the integer quotient and remainder in the form of a tuple, output(3,2)
print(divmod(20,6))
```

（3）当 z 不存在时，pow(x,y[,z])返回 x 的 y 次幂，当 z 存在时，返回 x 的 y 次幂再对 z 取余的结果，pow(x,y,z)函数在进行幂运算的同时可以进行模运算，比 x**y%z，先计算 x 的 y 次幂，再对 z 取余效率高。

(3)If z is not present, pow(x, y[, z]) returns x to the power of y, and if z is present, it returns the result of x to the power of y then modulus of z. The pow(x, y, z) function can perform both exponentiation and modulus operations simultaneously. It is more efficient than calculating x**y% z, as it first computes x to the power of y and then performs the modulus operation with z.

```
print(pow(3,2))             #Calculate 3², output 9
print(pow(3,2,4))           #3**2% 4, output 1
```

（4）round(number[,n]) 返回 number 舍入小数点后 n 位精度的值。如果 n 被省略，则返回最接近输入值的整数。

Python 中 round()函数采用的取舍算法为："四舍六入五考虑，五后非零就进一，五后为零看奇偶，五前为偶应舍去，五前为奇要进一"。

(4)round(number[,n]) returns the value of number rounded to the nearest n precision after the decimal point. If n is omitted, it returns the integer closest to the input value.

The round() function in Python follows the rounding algorithm: "round half to even", which means that when rounding a number ending in 5, if the digit before 5 is even, it should be rounded down; if it's odd, it should be rounded up to the nearest even number. This rule ensures that in cases where the digit after 5 is 0, the preceding digit is always even.

```
#3,return the integer closest to the input number
print(round(3.1415))
#-3,return the integer closest to the input number
print(round(-3.1415))
#4,return the integer closest to the input number
print(round(3.8415))
#3.13, round up if the digit after five is not zero
print(round(3.1250001,2))
#3.12, if the digit before five is even, it should be rounded down
print(round(3.125,2))
```

第2章　Python语法基础
Chapter 2　Basic Syntax of Python

```
#3.12, round up if the digit before five is odd
print(round(3.115,2))
```

因为二进制无法精确表示大多数浮点数，因此 Python 中浮点数无法保证完全精确，会导致部分数字取舍与期望不符。

Because binary cannot accurately represent most floating-point numbers, so floating-point numbers in Python cannot guarantee complete accuracy. This can lead to discrepancies between the expected and actual rounding of some numbers.

```
#0.1425 is stored as 0.14250000000000002 in the computer
print(round(3.1425,3))        #Expected output is 3.142, actual output is 3.143
print(round(2.675,2))         #Expected output is 2.68, actual output is 2.67
```

当 n 超过 number 小数位数，或 number 的小数部分都是 0 时，返回 number 的最短表示形式。Python 将 3.0000 和 3.0 认为是同一个对象，所以输出时会输出其最短表示形式 3.0。

When n exceeds the number of decimal places in the number, or when the decimal part of the number is all zeros, Python returns the shortest representation of the number. Python considers 3.0000 and 3.0 to be the same object, so when outputting, it will display the shortest representation as 3.0.

```
#Expected output is 3.00, actual output is the shortest representation of this floating-point number: 3.0
print(round(3.0000,2))
#Expected output is 3.1400, actual output is 3.14
print(round(3.14,4))
```

n 必须是整型数字，当 n 为浮点数时，会触发 TypeError。

The value of n must be an integer, and when n is a floating-point number, it will trigger a TypeError.

```
#TypeError:'float' object cannot be interpreted as an integer
print(round(1.25,2.0))
```

（5）max(arg1, arg2, …) 和 max(iterable) 从多个参数或一个可迭代对象中返回其最大值，有多个最大值时返回第一个。

（6）min(arg1, arg2, …) 和 min(iterable) 从多个参数或一个可迭代对象中返回其最小值，有多个最小值时返回第一个。

(5) max(arg1, arg2, …) and max(iterable) return the maximum value from multiple arguments or from an iterable object. If there are multiple maximum values, it returns the first one.

(6) min(arg1, arg2, …) and min(iterable) return the minimum value from multiple arguments or from an iterable object. If there are multiple minimum values, it returns the first one.

```
#Among the three integer objects 80, 100 and 1000, 1000 is the largest
print(max(80,100,1000))
#The list [49, 25, 88] is an iterable object, and the maximum value is 88
```

```
print(max([49,25,88]))
#Among the three integer objects 80, 100 and 1000, 80 is the smallest
print(min(80,100,1000))
#The list [49, 25, 88] is an iterable object, and the minimum value is 25
print(min([49,25,88]))
```

2.8 标准库使用
2.8 Standard Library Usage

模块是一个包含了 Python 代码的文件，通常以 .py 作为扩展名。它可以包含变量、函数、类等定义，并且可以被其他程序或模块导入和使用。模块提供了一种将代码组织成可重用单元的方式，通过将相关功能放在一个模块中，可以提高代码的可读性和维护性。模块具有独立性，可以在需要时单独导入和使用。

库则是由多个相关的模块组合而成的代码集合。库提供了一种封装了一定功能的软件组件，可以通过导入库来使用其中的模块和功能。库可以包含多个模块，并且这些模块之间可能存在某种关联或依赖关系。

库通常比模块更大、更全面，提供了更多的功能和工具，能够满足特定领域或任务的需求。

Python 库有两种：一种是标准库，另一种是第三方库。没有纳入标准库的库，需要在 Windows 的命令提示符或 Linux、macOS 的终端下使用以下命令安装后再使用，这样的库称为第三方库。Python 有众多优秀的第三方库，来帮助开发者快速完成功能开发，如 numpy 库、pandas 库等。注意以下命令是在操作系统的命令环境中运行，不能在

Module is a file containing Python code, typically with a .py extension. It can include variable, function, and class definitions, which can be imported and used by other programs or modules. Modules offer a way to organize code into reusable units, enhancing code readability and maintainability by grouping related functionalities into one module. Modules are self-contained and can be imported and utilized independently when needed.

Library is a collection of related modules combined into a code set. Libraries offer a way to package software components with specific functionalities, which can be utilized by importing the library and accessing its modules and features. A library can consist of multiple modules, and there may be certain relationships or dependencies among these modules.

Libraries are typically larger and more comprehensive than modules, offering a wider range of functionalities and tools to meet the requirements of specific domains or tasks.

Python libraries have two categories: the standard libraries and third-party libraries. Libraries that are not included in the standard library need to be installed using the following commands in the command prompt for Windows or terminal for Linux and macOS before they can be used. These libraries are referred to as third-party libraries. Python has a vast number of excellent third-party libraries to assist developers in quickly achieving various functionalities, such as numpy, pandas, etc. It is important to note that these commands are meant to be

Python 编程环境下运行。

run in the operating system's command environment and cannot be run within the Python programming environment.

```
pip install Module/Library Name
```

例如，安装 numpy 库使用以下命令：

For example, to install the numpy library, use the following command:

```
pip install numpy          #numpy is the name of a library that needs to be installed
```

Python 标准库是 Python 编程语言自带的、无须额外安装的库集合。这些库提供了大量的函数和类，用于执行各种常见的任务，从简单的文件操作到复杂的网络编程。使用标准库可以极大地简化 Python 程序的编写过程，并减少对第三方库的依赖。Python 标准库包含了许多不同的库，每个模块都专注于特定的功能或任务。本小节仅介绍 3 个模块：math 模块、string 模块和 random 模块。

The Python standard library is a collection of libraries that comes with the Python programming language and does not require additional installation. These libraries provide a wide range of functions and classes for performing various common tasks, from simple file operations to complex network programming. Using the standard library can greatly streamline the process of writing Python programs and reduce reliance on third-party libraries. The Python standard library includes many different libraries, each focusing on specific functionality or tasks. This section will introduce three modules: the math module, the string module, and the random module.

2.8.1 库和模块使用方法
2.8.1 How to Use Libraries and Modules

Python 中导入库（模块）的方法有两种，对于第三方库导入方法也一样。

There are two ways to import libraries(modules) in Python, and the same methods apply to importing third-party libraries as well.

(1) 使用关键字 import 导入库名，之后，程序可以调用库名中的所有函数，语法表示如下：

(1) Use the keyword import to import the library name. Then, the program can call all functions within the library. The syntax is as follows:

```
import <library_name>
```

这种方法使用库中常量或调用库中函数时，需要在前边加上库名和点，格式如下：

When using constants or calling functions from the library using this method, it is necessary to prefix the library name followed by a dot. The format is as follows:

```
library_name.function(or constant,variable)
```

（2）使用 from 和 import 关键字直接导入库中的函数，可以同时导入多个函数，各函数间用英文逗号分隔，也可以用通配符（＊）导入该库中的全部常量和函数，语法表示如下：

(2) Use the from and import keywords to directly import functions from the library. It is possible to import multiple functions simultaneously, separated by commas. Alternatively, one can import all constants and functions from the library using an asterisk(*). The syntax is as follows:

```
#Import multiple functions from a library, separated by commas
from <library_name> import <Function_Name1, Function_Name2,… Variable_Name>
#Import all functions from the library
from <library_name> import*        #* is a wildcard that represents all functions
```

这种方法导入库时，调用该库的函数不再需要指明函数所在库的名称。

When using this method to import a library, there is no need to specify the library name when calling functions from that library.

```
#Import the constant pi from math library
from math import pi
#Output the value of pi in math as 3.141592653589793
print(pi)
```

```
#Import all functions from math and directly reference the function names when calling them
from math import*
#Output the square root of 4 as 2.0
print(sqrt(4))
```

当编写的程序简单并且只需要使用到一个库，或者特定函数仅存在于一个库中时，可以选择直接导入整个库或仅导入所需的函数。在这种情况下，两种方式都是可行的。

然而，当编写的程序变得更为复杂并且涉及多个库时，情况就有所不同了。因为不同的库可能包含同名的函数，而这些同名函数的功能可能各不相同。为了避免在使用这些函数时产生混淆或错误，更好的做法是明确指定所需的函数来自哪个库。这样做可以确保代码的清晰性和准确性，防止因为函数名冲突而导致的问题。

When writing simple programs and only needing to use one library, or when the specific function exists only in one library, one can choose to either import the entire library or import only the required function. In this case, both ways are feasible.

However, when the program becomes more complex and involves multiple libraries, the situation is different. Because different libraries may contain functions with the same name, and these functions with the same name may have different functionalities. To avoid confusion or errors when using these functions, the better approach is to explicitly specify which library the required function comes from. This can ensure the clarity and accuracy of the code, preventing issues caused by function name conflicts.

2.8.2　math 模块
2.8.2　math Module

在数学运算中，除加、减、乘、除运算之外，还有更多其他的运算，如乘方、开方、三角函数运算、对数运算等。要实现这些运算，可以使用 Python 中的 math 模块，该模块提供了对 C 标准库中的数学函数的访问，包括常量、函数以及异常。

math 库提供了丰富的数学函数集，其中包括 21 个数论与表示函数，8 个幂和对数函数，9 个三角函数，6 个双曲函数，2 个角度转换函数，4 个特殊函数以及 5 个数学常数。

这些函数主要是对 C 标准库中的同名函数进行了简单的封装，因此它们主要支持整数和浮点数的运算，而不支持复数运算。若要在 Python 中进行复数运算，则应使用 cmath 模块，它提供了对复数进行数学运算的功能。

本章仅需掌握常数中的 pi，函数中的 fabs()、factorial()、gcd()、lcm()和 sqrt()即可。其他函数在需要时，可通过查阅官方文档学会使用，下面对部分常用的函数进行简单介绍。

（1）math.fabs(x)：以浮点数形式返回 x 的绝对值。

In mathematical operations, besides addition, subtraction, multiplication, and division, there are more other operations like exponentiation, square root, trigonometric functions, logarithmic operations, etc. To perform these operations, one can utilize the math module in Python, which grants access to mathematical functions from the C standard library, including constants, functions, and exceptions.

The math library provides a rich collection of mathematical functions, including 21 functions related to number theory and representation, 8 functions for power and logarithm, 9 trigonometric functions, 6 hyperbolic functions, 2 angle conversion functions, 4 special functions, and 5 mathematical constants.

These functions are mainly simple encapsulations of the same-named functions in the C Standard library, so they mainly support operations on integers and floating-point numbers, not complex number operations. If one wants to perform complex number operations in Python, one should use the cmath module, which provides functionalities for mathematical operations on complex numbers.

This section only requires mastery of the constant pi, and the functions fabs(), factorial(), gcd(), lcm(), and sqrt(). Other functions can be learned through consulting the official documentation when needed. Below is a simple introduction to some commonly used functions.

(1)math.fabs(x): Return the absolute value of x as a floating-point number.

```
import math
print(math.fabs(-5))        #Output 5.0
print(math.fabs(5))         #Output 5.0
```

（2）math.factorial(x)：返回 x 的阶乘，要求 x 为正整数，x 为负数或浮点数时则报错。

(2)math.factorial(x): Return the factorial of x, where x must be a positive integer. It will raise an error if x is a negative number or a float.

59

```
print(math.factorial(5))    #Output 120
```

（3）math.gcd(*integers)：返回给定的整数参数的最大公约数。如果参数之一非0，则返回值将是能同时整除所有参数的最大正整数。如果所有参数为0或无参数，则返回值为0。在Python 3.9版后增加了对任意数量的参数的支持，Python 3.8或之前版本只支持两个参数。

（4）math.lcm(*integers)：返回给定的整数参数的最小公倍数。如果所有参数均非0，则返回值将是为所有参数的整数倍的最小正整数。如果参数之一为0，则返回值为0。不带参数的lcm()返回1。此函数为Python 3.9版新增函数。

(3) math.gcd(*integers): Return the greatest common divisor of the given integer parameters. If at least one of the parameters is not 0, the return value will be the largest positive integer that can integer divide all the parameters simultaneously. If all the parameters are 0 or there are no parameters, the return value is 0. Support for an arbitrary number of parameters was added in Python 3.9, Python 3.8 and earlier versions only support two parameters.

(4) math.lcm(*integers): Return the least common multiple of the given integer parameters. If all the arguments are not 0, the return value will be the smallest positive integer that is a multiple of all the arguments. If any of the parameters is 0, the return value is 0. The lcm() function without parameters returns 1. This function was added in Python 3.9.

```
print(math.gcd(55,44,33))      #Output 11
print(math.gcd(0,0))           #Output 0
print(math.lcm(33,11,3))       #Output 33
print(math.lcm())              #Output 1
```

（5）math.sqrt(x)：返回x的平方根，结果为浮点数。

(5) math.sqrt(x): Return the square root of x, the result is a floating-point number.

```
import math
print(math.sqrt(2))   #1.4142135623730951
print(math.sqrt(3))   #1.7320508075688772
```

2.8.3 string 模块
2.8.3 string Module

Python 标准库中包含了一个名为 string 的模块，它提供了一系列与字符串相关的常量和类。这些常量和类在字符串处理中非常有用，尤其是当需要一组特定的字符或字符串时。使用 string 模块时，需先导入该模块（import string），string 模块定义的常量如表 2-11 所示。

The Python standard library includes a module called string that provides a range of constants and classes related to strings. These constants and classes are very useful in string processing, especially when one needs a specific set of characters or strings. When using the string module, one needs to first import the module (import string). The constants defined in the string module are as shown in Table 2-11.

表 2-11 string 模块定义的常量
Table 2-11　Constants Defined in the string Module

字符串常量 String Constants	字符集 Character Set
string.ascii_letters	'abcdefghijklmnopqrstuvwxyzABCDEFGHIJKLMNOPQRSTUVWXYZ'
string.ascii_lowercase	'abcdefghijklmnopqrstuvwxyz'
string.ascii_uppercase	'ABCDEFGHIJKLMNOPQRSTUVWXYZ'
string.digits	'0123456789'
string.hexdigits	'0123456789abcdefABCDEF'
string.octdigits	'01234567'.
string.punctuation	'!"#$%&\'()*+,-./:;<=>?@[\\]^_`{|}~'
string.printable	'0123456789abcdefghijklmnopqrstuvwxyzABCDEFGHIJKLMNOPQRSTUVWXYZ!"#$%&\'()*+,-./:;<=>?@[\\]^_`{|}~ \t\n\r\x0b\x0c'
string.whitespace	'\t\n\r\x0b\x0c'

```
import string
#Output: 'abcdefghijklmnopqrstuvwxyzABCDEFGHIJKLMNOPQRSTUVWXYZ'
print(string.ascii_letters)
#Output: '0,1,2,3,4,5,6,7,8,9'
print(','.join(string.digits))
#Output: '!"#$%&\'()*+,-./:;<=>?@[\\]^_`{|}~'
print(string.punctuation)
#Output: ' \t\n\r\x0b\x0c'(Here contains spaces, tabs, line breaks, and so on)
print(string.whitespace)
```

2.8.4　random 模块
2.8.4　random Module

随机数是在一定范围内产生的看似无规律、无法预测的数字或数字序列。真正的随机数应该是完全不可预测的，并且在每次生成时都是独立的。随机数在统计学、密码学等领域有非常广泛的应用。真正的随机数是使用物理方法产生的，如掷钱币、骰子、转轮、使用电子元器件时的噪声、核裂变等。这样的随机数发生器叫作物理性随机数

Random numbers are seemingly irregular and unpredictable numbers or number sequences generated within a certain range. True random numbers should be entirely unpredictable and independent each time they are generated. Random numbers have a wide range of applications in statistics, cryptography, and other fields. True random numbers are generated using physical methods such as flipping coins, rolling dice, spinning wheels, noisewhen using electronic component, nuclear fission, etc. These types of random number generators are known as

发生器，它们的缺点是技术要求比较高。

　　计算机随机数是在计算机程序中生成的看似随机的数字序列。这些数字实际上是通过特定的算法生成的，因此并不是真正的随机数，而是被称为伪随机数。伪随机数生成器（PRNG）是一种常用的生成随机数的方法，它通过一个初始值（种子）作为输入，经过一系列的计算生成一串数字序列。这个序列看起来是随机的，但实际上是由初始种子和算法完全确定的。

　　Python 的 random 模块是一个提供各种随机数生成功能的库。它包含了多种生成伪随机数的方法，可以用于模拟、统计分析、游戏开发等多种场景。"种子"是这个算法开始计算的第一个值，如果随机数种子一样，那么后续所有"随机"结果和顺序也都是完全一致的。

　　在 Python 中，当不设置随机数种子时，解释器会使用系统时间作为种子，使每次生成的随机数不同。当希望得到的随机数据可预测时，可以设置使用相同的种子，使后续产生的随机数相同。

　　在 Python 中，random 是一个标准库，使用随机数函数时，使用 import 关键字或使用 from 和 import 关键字直接导入 random 模块即可。

physical random number generators. One downside is that they require high technical requirements.

　　Computer random numbers are seemingly random number sequences generated within computer programs. These numbers are actually generated through specific algorithms, so they are not truly random numbers, but rather known as pseudo-random numbers. Pseudo-random numbers generators (PRNG) are a common method for generating random numbers. It takes an initial value(seed) as input and through a series of calculations generate a sequence of numbers. This sequence appears random, but it is actually completely determined by the initial seed and the algorithm used.

　　The random module in Python is a library that provides various random number generation functions. It includes multiple methods for generating pseudo-random numbers that can be used in simulations, statistical analysis, game development, and many other scenarios. The "seed" is the first value from which the algorithm begins calculations. If the random number seed is the same, then all subsequent "random" results and sequences will be completely consistent.

　　In Python, if a random number seed is not set, the interpreter uses the system time as the seed, ensuring that each generated random number is different. When predictable random data is desired, setting the same seed will produce the same random numbers in subsequent iterations.

　　In Python, random is a standard library. When using random number functions, one can either use the "import" keyword or directly import the random module using the "from" and "import" keyword.

```
import random
#from random import*
```

　　random 模块包含了一系列函数，可提供多种形式的随机数序列，random 模块中主要函数如表 2-12 所示。

　　The random module contains a range of functions that provide various forms of random number sequences. The main functions of the random module are described in Table 2-12.

第2章　Python语法基础
Chapter 2　Basic Syntax of Python

表2-12　random 模块中主要函数
Table 2-12　The Main Functions of the random Module

函数 Function	描述与示例	Description and Examples
random.seed (a=None,version=2)	初始化随机数生成器,如果参数 a 被省略或为 None,则使用系统时间做种子。seed 必须是下列类型之一：NoneType、int、float、str、bytes 或 bytearray random.seed(20) #用整数20做种子	Initialize the random number generator. If the parameter 'a' is omitted or is None, the system time is used as the seed. The seed must be one of the following types：NoneType, int, float, str, bytes, or bytearray. random.seed(20) #Use the integer 20 as the seed
random.randint(a,b)	产生[a, b]之间(包括 b)的一个随机整数 print(random.randint(1, 10)) #输出9	Generate a random integer in [a, b] (including b). Print(random.randint(1, 10)) #Output 9
random.random()	产生[0.0, 1.0)之间的一个随机浮点数 print(random.random()) #输出 0.8432521696967962	Generate a random float number in [0.0, 1.0). print(random.random()) #Output 0.8432521696967962
random.uniform(a,b)	产生[a, b)之间的一个随机浮点数 print(random.uniform(5.5, 10.0)) #输出 6.487385989202917	Generate a random float number in [a, b). print(random.uniform(5.5, 10.0)) #Output 6.487385989202917

random.randrange(stop)
random.randrange
(start,stop[,step])
	从[0, stop)(不含 stop)中随机产生一个整数
	从[start, stop)、步长为 step 的序列中随机产生一个整数
	print(random.randrange(10))
	#输出2
	print(random.randrange(0, 10, 2))
	#输出6
		Generate a random integer in [0, stop) (excluding stop). Generate a random integer from a sequence [start, stop) with a step of step.
		print(random.randrange(10))
		#Output 2
		print(random.randrange(0, 10, 2))
		#Output 6

random.choice(seq)	从非空序列 seq 中随机产生一个元素,当序列为空时,触发异常 print(random.choice(['win', 'lose', 'draw'])) #输出 draw	Generate a random element from the non-empty sequence seq, trigger an exception when the sequence is empty. print(random.choice(['win', 'lose', 'draw'])) # Output draw

```
import random
#Generate a random floating-point number between 0 and 1
print(random.random())
#Generate a random integer between 1 and 10
print(random.randint(1, 10))
```

```
#Randomly select an element from a list
my_list=[1, 2, 3, 4, 5]
print(random.choice(my_list))
#Shuffle the order of elements in a list
random.shuffle(my_list)
print(my_list)
#Randomly select 3 non-repeating elements from a list
print(random.sample(my_list, 3))
#Set a random seed to ensure the same random number sequence is generated every time the code runs
random.seed(123)
#Using seed 123, the same random number will be obtained each time
print(random.random())
```

输出： | Output:

```
0.4950042380455387
8
1
[4, 3, 2, 5, 1]
[1, 2, 5]
0.052363598850944326
```

2.9 字符串操作
2.9 String Operations

2.9.1 索引
2.9.1 Index

在 Python 中，字符串索引用于访问字符串中的特定字符，字符串中的每个字符都有一个索引。所谓的索引是指通过字符串的序号返回其对应字符值的操作。Python 提供了两套索引：正向索引和逆向索引。

(1)正向索引从 0 开始，终止值为序列长度减 1(元素个数减 1，即 len(s)-1)。

(2)逆向索引从-1 开始，终止值为负的序列长度(即-len(s))。

In Python, string indexing is used to access specific characters in a string, each character in the string has an index. The index refers to the operation of returning the corresponding character value by the sequence number of the string. Python provides two sets of indexing: forward indexing and reverse indexing.

(1)Forward indexing starts from 0 and ends at the sequence length minus 1(number of elements minus 1, i.e., len(s)-1).

(2)Reverse indexing starts from -1 and ends at the negative sequence length(i.e., -len(s)).

第2章　Python语法基础
Chapter 2　Basic Syntax of Python

两种索引可以单独使用一种，也可以混合使用，结合两种表示方法可以方便地对字符串进行索引和切片。图2-1中给出正向和逆向两种索引编号规则的示例。对于字符串，英文字母、汉字、空格和各种符号都各占一个字符位。

索引的方法如下：

字符串名[序号]

The two types of indexing can be used separately or combined, and combining both representations makes it easy to index and slice strings. Example of both forward and reverse indexing rules is provided in Figure 2-1. For strings, English letters, Chinese characters, spaces, and various symbols each occupy one character position.

The method for indexing is as follows:

string_name[index]

正向索引 Forward	0	1	2	3	4	5	6	7	8	9	10	11
字符串 String	H	e	l	l	o		P	y	t	h	o	n
逆向索引 Reverse	-12	-11	-10	-9	-8	-7	-6	-5	-4	-3	-2	-1

图 2-1　字符串索引

Figure 2-1　String Indexing

字符串的元素可以按正向序号进行索引或按逆向序号进行索引，通过序号获取对应元素，索引序号必须为整数，不可为浮点数。另外，需要注意的是，当使用的索引值超出字符串索引值范围时，将会产生"索引超出范围"的错误。例如，试图用s[13]获取字符串s中不存在的字符值时，会得到"IndexError: string index out of range"的出错提示。

The elements of a string can be indexed using either forward or reverse indexing, and the corresponding element can be accessed using the index number. The index number must be an integer and cannot be a floating-point number. Additionally, it is important to note that using an index value beyond the range of the string will result in an "index out of range" error. For example, trying to access a character value in string "s" with s[13] when the character does not exist in the string will prompt an error message "IndexError: string index out of range".

```
s='Hello Python!'
#Forward indexing by index, return the character 'o' at index 4
print(s[4])
#Reverse indexing by index, return the last character '!'
print(s[-1])
#IndexError: string index out of range
print(s[13])
```

2.9.2　切片
2.9.2　Slice

切片用于从序列类型（如字符串、列表、元组等）中提取部分元素。切片语法允许指定开始和结束

Slice is used to extract part of the elements from sequence types(such as strings, lists, tuples, etc.). Slice syntax allows one to specify a start and end index(optional),

65

索引(可选的)，以及一个步长(也是可选的)，来创建原始序列的一个子集。字符串支持切片操作，切片的方法如下。

as well as a step(also optional), to create a subset of the original sequence. Strings support slicing operations, and the method of slicing is as follows.

> seq[start:end:step]

seq 是字符串的名字。切片操作涉及 3 个关键参数：start、end 和 step。通过合理设置这 3 个参数，可以灵活地从 seq 字符串中提取所需的子字符串。

start 指的是切片开始的位置(包含该位置元素)，即想要从哪个字符开始提取。默认为 0，也就是字符串的第一个字符。如果切片是从字符串的第一个字符开始，那么可以省略 start。

end 表示切片结束的位置(不包含该位置元素)，即想要提取到哪个字符为止，默认为序列的长度。如果切片是到字符串的最后一个字符，那么可以省略 end。

step 为步长，用于决定了取值的间隔。默认情况下，step 是 1，意味着每次取相邻的字符。但也可以设置 step 为其他正数或负数。正数表示按照指定的间隔向字符串的末尾方向取值，而负数则表示向字符串的开头方向取值。需要注意的是，step 不能设置为 0，这会导致无法取值。

在切片 seq[start:end:step] 中，是包括值 seq[start]，而不包括值 seq[end] 的。

如果切片结果需要包含最后一个元素(len(s)-1)，则结束位置的序号 end 应该设为 len(s) 或省略结束位置序号，即应该使用切片 seq[start:len(s)] 或 seq[start:]。

seq is the name of the string. The slicing operation involves three key parameters: start, end, and step. By setting these three parameters reasonably, one can flexibly extract the desired substring from the seq string.

start refers to the position where the slice starts(including the element at that position), in other words, where one wants to begin extracting from. The default is 0, which is the first character of the string. If the slice starts from the first character of the string, the start can be omitted.

end indicates the position where the slice ends(excluding the element at that position), in other words, where one wants to extract up to. The default is the length of the sequence. If the slice goes up to the last character of the string, the end can be omitted.

step is the step that determines the interval of values taken. By default, the step is 1, meaning that each time adjacent characters are taken. However, one can also set the step to other positive or negative numbers. A positive number indicates extracting values towards the end of the string at the specified interval, while a negative number means extracting values towards the beginning of the string. It is important to note that the step cannot be set to 0, as this would result in an inability to extract values.

In the slice seq[start:end:step], the value seq[start] is included, but the value seq[end] is not included.

If one wants to include the last element(len(s)−1) in the slice result, the index of the end position should be set to len(s) or omitted. Therefore, one should use the slice seq[start:len(s)] or seq[start:] to include the last element.

第2章　Python语法基础
Chapter 2　Basic Syntax of Python

```
s='Hello Python!'
#Slice according to the index [6:8], output the characters 'Py' excluding the character at the ending index
print(s[6:8])
#Slice from the start to index 5, excluding 5, output 'Hello'
print(s[:5])
#Slice from index 6 to the end of the string, output 'Python!'
print(s[6:])
#Use negative indexing, exclude the right boundary element, output 'on'
print(s[-3:-1])
#Mix positive and negative indexing, output 'Python'
print(s[6:-1])
#Slice the string from the beginning to the end, output 'Hello Python!'
print(s[::])
#Slice with a step of -1, output '!nohtyP olleH'
print(s[::-1])
#With a step of 2, output the elements at even indexes, output 'HloPto!'
print(s[::2])
```

2.9.3　拼接与重复
2.9.3　Concatenation and Repetition

当"+"号两侧数据类型都是数值型时，完成的是数学上的加法运算，而当两侧数据类型都是字符串类型时，完成的是字符串拼接功能。拼接是通过"+"号将两个字符串拼接为一个新字符串，新字符串包含两个字符串中所有元素。注意，"+"号两侧数据类型不能一侧是数值类型，另一侧是字符串类型。

当"*"号两侧数据类型都是数值型时，完成的是数学上的乘法运算，而当一侧是字符串类型，另一侧是整数类型时，完成的是字符串重复操作。s*n是将一个字符串s乘以一个整数n产生一个新字符串，新字符串是s中的元素重复n次。当n小于或等于0时会被当作0来处理，此时序列重复0次的操作将产生一个空序列。

When the data types on both sides of the "+" sign are numerical, it performs mathematical addition, while when both sides are strings, it performs string concatenation. Concatenation is done by using the "+" sign to combine two strings into a new string that contains all elements from both original strings. Note that the data types on both sides of the "+" sign cannot be mixed, i.e., one side being numerical and the other side being a string.

When the data types on both sides of the "*" sign are numerical, it performs multiplication in mathematics, while when one side is a string and the other side is an integer, it performs string repetition. s*n means multiplying a string s by an integer n to generate a new string, where the new string consists of elements of s repeated n times. When n is less than or equal to 0, it is treated as 0, resulting in an empty sequence when the repetition operation is done 0 time.

```
year=2024
s='year'
#Repeat the string 10 times
print('='*10)
#String concatenation, integers involved in concatenation need to be converted to strings first
print(str(year)+s)
print('='*10)
```

输出： | Output:

```
==========
2024year
==========
```

【例2-3】输出身份证信息。

中国的居民身份证号是一个有18个字符的字符串，其各位上的字符代表的意义如下。

第1、2位数字表示所在省份的代码，例如，辽宁省的省份代码是21。

第3、4位数字表示所在城市的代码，例如，鞍山市的代码是03。

第5、6位数字表示所在区县的代码，例如，铁东区的代码是02。

第7~14位数字表示出生年、月、日。

第15、16位数字表示在同一地址码所标识的区域范围内，对同年、同月、同日出生的人编定的顺序号。

第17位数字表示性别，奇数表示男性，偶数表示女性。

第18位数字是校检码，用来检验身份证号的正确性。校检码可以是0~9中的一个数字，也可以是字母X。

输入一个身份证号，输出其出

[Example 2-3] Outputting identity card information.

The resident identity card number in China is a string of 18 characters, and the meanings of each digit are as follows.

The first and second digits represent the code of the province of residence. For example, the province code for Liaoning is 21.

The third and fourth digits represent the code of the city. For example, the code for Anshan city is 03.

The fifth and sixth digits represent the code of the district. For example, the code for Tiedong district is 02.

The seventh to fourteenth digits represent the date of birth(year, month, day).

The fifteenth and sixteenth digits represent the sequential number assigned to people born on the same year, month, and day in the area identified by the same address code.

The seventeenth digit represents the gender, odd for male and even for female.

The eighteenth digit is the check digit used to verify the correctness of the identity card number. The check digit can be a number from 0 to 9 or the letter X.

Input an identity card number and output the date of

生年月日。（注：本书测试所用身份证号是用程序模拟生成的虚拟号码。）

用字符串切片的方法获取身份证号码中代表出生年月日的子串，用"+"连接后输出。

birth.(Note: The identity card numbers used for testing in this book are simulated and virtual numbers generated by the program.)

Use the string slicing method to obtain the substring representing the date of birth from the identity card number, and then output the connected substring with "+".

```
#Extract birthday from ID card, involving knowledge of string length, slicing, concatenation, and so on
id_number=input('Input the ID number: ')
#Obtain the year at index 6 to 9 from the string ID
year=id_number[6:10]
#Obtain the month at index 10 and 11 from the string ID
month=id_number[10:12]
#Obtain the day at index 12 and 13 from the string ID
day=id_number[12:14]
#String concatenation
print('Born in: '+ year+' year '+month+' month '+day+' day')
#Output using the f-string method
#print(f'Born in:{year} year {month} month {day} day')
```

2-3

输入： | Input:

2103022024010111121

输出： | Output:

Born in: 2024 year 01 month 01 day

通过字符串切片，用 id_number[6:10]、id_number[10:12]、id_number[12:14]分别获取出生年份、月份和日期。在切片时，切分出来的子字符串包括左边界，但不包括右边界。

语句 print('Born in: '+ year+' year '+month+' month '+day+' day')中，采用6个"+"将4个字符串和3个字符串变量拼接成一个新的字符串并输出。这里也可以使用 str.format()方式或使用"f"前缀格式化字符串输出，这两种方法不限

Through string slicing, extract the birth year, month, and day using id number[6:10], id number[10:12], and id number[12:14] respectively. When slicing, the substring obtained includes the left boundary but does not include the right boundary.

In the statement "print('Born in: '+year+' year'+month+' month'+day+' day')", using 6 "+" signs to concatenate 4 strings and 3 string variables into a new string and output it. Here, one can also use the str.format() method or the "f" prefix to format string output, which do not restrict variable types and more convenient to use.

制变量类型，使用更方便些。

【例2-4】温度转换。温度的表示有摄氏度和华氏度两种方式。编写程序完成以下功能：将用户输入的华氏度转换为摄氏度，输入的摄氏度转换为华氏度。转换公式如下（C 表示摄氏度、F 表示华氏度）：

[Example 2-4] Temperature conversion. The temperature can be expressed in two ways: Celsius and Fahrenheit. Write a program to achieve the following functions: the user inputs Fahrenheit to convert to Celsius and inputs Celsius to convert to Fahrenheit. The conversion formulas are as follows(C represents Celsius, F represents Fahrenheit):

$$C=(F-32)/1.8$$
$$F=C\times1.8+32$$

要求输入输出的摄氏度单位用大写字母"C"或小写字母"c"，华氏度单位用大写字母"F"或小写字母"f"。温度可以是整数或小数，例如：12.34C 指摄氏度 12.34 度；87.65F 指华氏度 87.65 度。输出转换后的温度，保留小数点后 2 位小数。对于输入的数据 12.34C，可以用字符串切片方法"12.34C"[-1]获取到最后一位字符"C"，再通过分支结构根据最后一位字符进行相应的转换。

The requirement is to input and output the temperature in Celsius using the uppercase letter "C" or lowercase letter "c", and in Fahrenheit using the uppercase letter "F" or lowercase letter "f". The temperature can be an integer or a decimal number, for example: 12.34C refers to 12.34 degrees Celsius; 87.65F refers to 87.65 degrees Fahrenheit. Output the converted temperature with 2 decimal places. For the input data 12.34C, one can use the string slicing method "12.34C"[-1] to retrieve the last character "C", then use a conditional structure to perform the corresponding conversion based on the last character.

```python
#Enter the temperature value, followed by the temperature unit symbol
temperature=input('Input the temperature: ')
#Whether the symbol exists in 'Ff'
if temperature[-1] in 'Ff':
    #Calculate using the Fahrenheit to Celsius formula
    C=(float(temperature[:-1])-32)/1.8
    #Output the converted temperature and unit
    print("{:.2f}C".format(C))
#Whether the symbol exists in 'Cc'
elif temperature[-1] in 'Cc':
    #Calculate using the Celsius to Fahrenheit formula
    F=1.8*float(temperature[:-1])+32
    #Output the converted temperature and unit
    print("{:.2f}F".format(F))
#The last character is not a character in "Cc" or "Ff"
else:
```

```
#Output error message
    print('Data error!')
```

输入： | Input:

```
88.81F
```

输出： | Output:

```
31.56C
```

输入： | Input:

```
31.56C
```

输出： | Output:

```
88.81F
```

2.9.4 常用字符串处理方法
2.9.4 Common String Processing Methods

Python 提供了许多用于处理字符串的内置方法，这里只介绍一些常用的方法，如表 2-13 所示。

strip()、lstrip()和 rstrip()：去除字符串两侧的空白字符（包括空格、制表符、换行符等），或只去除左侧或右侧的空白字符。例如：

Python provides many built-in methods for handling strings, here are some commonly used methods, as shown in Table 2-13.

strip(), lstrip() and rstrip(): Remove whitespace characters (including spaces, tabs, newline characters, etc.) from both sides of a string, or from only the left side or right side. For example:

```
s1='hello python'
s2='    hello world    '
print(s1.upper())
print(s1.find('p'))
print(','.join(s1))
print()
print(s2.strip())      #Output: "hello world"
print(s2.lstrip())     #Output: "hello world    "
print(s2.rstrip())     #Output: "    hello world"
```

输出： | Output:

```
HELLO PYTHON
6
h,e,l,l,o, ,p,y,t,h,o,n
```

```
hello world
  hello world
    hello world
```

表 2-13 常用字符串处理函数方法
Table 2-13 Common String Processing Methods

方法名 Method Name	描述	Description
str.upper()/str.lower()	转换字符串 str 中所有字母为大写/小写	Convert all letters in the string 'str' to uppercase/lowercase
str.strip()	用于移除字符串开头、结尾指定的字符(参数省略时去掉空白字符,包括\t、\n、\r、\x0b、\x0c)	Used to remove specified characters at the beginning and end of a string (when the parameter is omitted, remove whitespace characters, including \t, \n, \r, \x0b, and \x0c)
str.join(iterable)	以字符串 str 作为分隔符,将可迭代对象 iterable 中的字符串元素拼接为一个新的字符串。当 iterable 中存在非字符串元素时,返回一个 TypeError 异常	Use the string str as a separator, concatenate the string elements in the iterable object iterable into a new string. When there are non-string elements in iterable, it raises a TypeError exception
str.split(sep=None, maxsplit=-1)	根据分隔符 sep 将字符串 str 切分成列表,sep 参数省略时根据空格切分,可指定逗号或制表符等为分隔符。maxsplit 值存在且非-1 时,最多切分 maxsplit 次	Split the string str into a list based on the separator sep. When the sep parameter is omitted, it will split based on spaces by default, but one can also specify a comma or a tab as the separator. If the maxsplit value is present and not -1, it will split at most maxsplit times
str.count(sub[, start[, end]])	返回 sub 在字符串 str 中出现的次数,如果指定了 start 或者 end,则返回指定范围内 sub 出现的次数	Return the number of occurrences of sub in the string str, if start or end is specified, return the number of occurrences of sub within the specified range
str.find(sub[, start[, end]])	检测 sub 是否包含在字符串 str 中,如果是,则返回开始的索引值,否则返回-1。如果指定了 start 和 end,则检查 sub 是包含在指定范围内	Check if sub is contained in the string str, if yes, return the starting index value, otherwise return -1. If start and end are specified, check if it is contained within the specified range
str.replace(old, new[, count])	把字符串 str 中的 old 替换成 new,如果 count 指定,则替换不超过 count 次,否则有多个 old 子串时全部替换为 new	Replace old with new in the string str, if count is specified, replace no more than count times. Otherwise, replace all occurrences of old substring with new when there are multiple occurrences

续表

方法名 Method Name	描述	Description
str.index(sub[, start[, end]])	与find()方法一样，返回子串存在的起始位置，如果sub在字符串str中不存在，则抛出一个异常	Like the find() method, return the starting position where the substring exists. If sub does not exist in the string str, an exception is raised
for <var> in <string>	对字符串string进行遍历，依次将字符串string中的字符赋值给前面的变量var	Iterate through string and assign each character in the string to the preceding variable var one by one

2.10 本章小结
2.10 Summary of This Chapter

本章介绍了 Python 语法基础知识，包括关键字、标识符以及命名规范，对象和变量，通过赋值表达式将变量和对象绑定，数值类型、字符串类型和它们之间的相互转换，运算符和表达式，常用的内置数学函数，标准库的使用和字符串操作等。具体内容如下。

（1）关键字和标识符：Python 对于标识符的命名规范要求为由字母、数字和下画线组成，首字符不能是数字，关键字也符合标识符命名规范。

（2）对象和变量：Python 中通过赋值符号（=）将变量和对象绑定，通过变量可以操作所绑定的对象。

（3）数值类型：包括了整数类型、浮点数类型和复数类型。

（4）字符串类型：Python 中字符串可以由一对单引号、一对双引号或者一对三引号来表示。

（5）常用内置数学函数和常用标准库，包括 math 模块、string 模

This chapter introduces the basic knowledge of Python syntax, including keywords, identifiers and naming conventions, objects and variables, binding variables and objects through assignment expressions, numerical types, string types and their mutual conversion, operators and expressions, commonly used built-in mathematical functions, the use of standard libraries, and string operations. The specific contents are as follows.

(1)Keywords and identifiers: Python requires the naming convention for identifiers to consist of letters, numbers, and underscores, with the first character not being a number. Keywords also conform to the identifier naming convention.

(2)Objects and variables: In Python, variables are bound to objects through the assignment operator (=), and objects bound to variables can be manipulated through the variables.

(3)Numerical types: This includes integer types, floating-point types, and complex types.

(4)Strings in Python can be represented by a pair of single quotes, a pair of double quotes, or a pair of triple quotes.

(5)Commonly used built-in mathematical functions and common standard libraries, including the the math

块、random 模块。

(6)字符串的索引、切片、拼接和重复操作和字符串常用的处理方法。

module, string module, and random module.

(6)String indexing, slicing, concatenation, and repetition operations, as well as commonly used string manipulation methods.

2.11 本章练习
2.11 Exercise for This Chapter

2.1 编写程序，输入一个数字作为圆的半径，计算并输出这个圆的面积，π 值取 3.14，输出保留小数点后 2 位数字。

2.2 编写程序，输入两个数字 a 和 b，计算并输出这两个数的和、差、积、商。

2.3 用户输入用逗号分隔的 3 个数字，输出其中数值最大的一个。

2.4 用户输入用逗号分隔的多个数字，输出其中数值最小的一个的绝对值。

2.5 用户在同一行中输入用逗号分隔的两个正整数 a 和 b，以元组形式输出 a 除以 b 的商和余数。

2.6 在两行中分别输入一个正整数 M、N，在一行中依次输出 M 和 N 的最大公约数和最小公倍数，两数字间以 1 个空格分隔。

2.7 根据下面公式计算并输出 x 的值，a 和 b 的值由用户在两行中输入，括号里的数字是角度值，要求圆周率的值使用数学常数 math.pi，三角函数的值用 math 库中对应的函数进行计算。请编程计算并输出表达式的值。

2.1 Write a program that takes a number as the radius of a circle, calculates and outputs the area of the circle. Take the value of π as 3.14 and output the result rounded to two decimal places.

2.2 Write a program that takes two numbers, "a" and "b" as input and calculates the sum, difference, product, and quotient of these two numbers. Then output the results.

2.3 The user inputs three numbers separated by commas, and the program outputs the largest of the three numbers.

2.4 The user inputs multiple numbers separated by commas, and the program outputs the absolute value of the smallest number among them.

2.5 The user inputs two positive integers, "a" and "b", separated by a comma on the same line. The program outputs the quotient and remainder when "a" is divided by "b" as a tuple.

2.6 Input a positive integer "M" on the first line and a positive integer "N" on the second line. Output the greatest common divisor and least common multiple of "M" and "N" on the same line, separated by a single space.

2.7 Calculate and output the value of "x" based on the following formula. The values of "a" and "b" are entered by the user on two separate lines. The numbers in parentheses represent angle values, and the mathematical constant "pi" should use the constant math.pi. The trigonometric functions should be calculated using the corresponding functions in the "math" library. Please program the calculation and output the value of the expression.

$$x = \frac{-b + \sqrt{2a\sin(60)\cos(60)}}{2a}$$

2.8 BMI 指数，即身体质量指数，是用体重(单位：kg)除以身高(单位：m)的平方得出的数字，是目前国际上常用的衡量人体胖瘦程度的一个标准。BMI 中国标准为：

分类	BMI 值范围
偏瘦：	<=18.5
正常：	18.5~24.0
过重：	24.0~28.0
肥胖：	>=28.0

如果用户的 BMI 值大于 24.0 或小于 18.5，则可能需要多加锻炼身体。编写程序，用户输入身高和体重，计算对应的 BMI 指数。

2.8 BMI, or Body Mass Index, is a number calculated by dividing a person's weight in kilograms by the square of his/her height in meters. It is an internationally recognized standard for assessing body weight. The BMI values in the Chinese standard classification are as follows:

Classification	BMI value
Underweight:	<=18.5
Normal weight:	18.5~24.0
Overweight:	24.0~28.0
Obese:	>=28.0

If one's BMI is greater than 24.0 or less than 18.5, one may need to exercise more. Write a program where the user inputs the height and weight to calculate the corresponding BMI.

本章练习题参考答案

第 3 章 控制结构
Chapter 3 Control Structure

控制结构是构建程序逻辑的基础，主要包括 3 种类型：顺序结构、分支结构和循环结构。顺序结构按照代码的顺序逐行执行代码，是最基本的程序结构。分支结构根据条件判断的结果，选择执行不同的代码块，这种结构使得程序能够根据不同的情况作出不同的响应，增强了程序的灵活性和适应性。循环结构则用于重复执行某段代码，这在处理重复任务时非常高效。

The control structure is the foundation for constructing program logic. It mainly consists of three types: sequential structure, conditional structure and loop structure. Sequential structure executes code line by line in the order they appear, serving as the most basic program structure. Conditional structure selects different code blocks to execute based on the result of a condition, enhancing the flexibility and adaptability of programs to respond differently in different situations. Loop structure allows for repeating a segment of code, making it highly efficient for handling repetitive tasks.

3.1 顺序结构
3.1 Sequential Structure

Python 的顺序结构指的是程序按照代码的书写顺序，从上到下、从左到右依次执行代码的一种基本控制结构。在顺序结构中，程序会严格按照语句的排列顺序执行，不会跳过任何语句，也不会重复执行任何语句。下面是一个简单的顺序结构的例子，演示如何按照顺序执行一系列语句来计算两个数的和。

The sequential structure in Python refers to the basic control structure where the program executes the statements in the order they are written, from top to bottom and from left to right. In the sequential structure, the program strictly follows the arrangement of statements, without skipping or repeating any statement. Here is a simple example of sequential structure that demonstrates how to calculate the sum of two numbers by executing a series of statements in order.

```
#Define two variables
a=5
b=3
```

第3章　控制结构
Chapter 3　Control Structure

```
#Calculate the sum of two numbers
sum=a+b
#Output results
print("The sum of", a, "and", b, "is", sum)
```

在这个例子中，首先定义了两个变量 a 和 b，分别赋值为 5 和 3。接着，计算这两个数的和，将结果存储在变量 sum 中。最后，使用 print()函数输出 a+b 的和。当运行这段代码时，Python 解释器会按照顺序执行每一条语句，最终输出下面的结果：

In this example, two variables, a and b, are defined with values 5 and 3 respectively. Then, the sum of these two numbers is computed and stored in a variable named "sum". Finally, the result of "a+b" is printed using the print() function. When this code is executed, the Python interpreter will execute each statement in order, resulting in the following output:

```
The sum of 5 and 3 is 8
```

顺序结构是编程中最基础、最直观的控制结构，也是构建更复杂逻辑的基础，其流程图如图 3-1 所示。

The sequential structure is the most basic and intuitive control structure in programming, serving as the foundation for building more complex logic. Its flowchart is illustrated in Figure 3-1.

图 3-1　顺序结构流程图
Figure 3-1　Sequential Structure Flowchart

【例 3-1】使用顺序结构来计算圆的面积和周长，首先需要定义圆的半径，然后使用圆的面积和周长的数学公式来进行计算。π 是一个常数，通常近似值为 3.14159。计算并输出给定半径的圆的面积和周长。

[Example 3-1] To calculate the area and circumference of a circle using sequential structure, we first need to define the radius of the circle. Then, we can use the mathematical formulas to calculate the area and circumference of a circle. The value of π is a constant and is usually approximated as 3.14159. Calculate and output the area and circumference of a circle with a given radius.

77

```
#Import the math module to use mathematical functions such as π and square root
import math
#Define the radius of the circle
radius=5                          #Change this value to any radius you want
area=math.pi*(radius**2)          #Calculate the area of a circle
circumference=2*math.pi*radius    #Calculate the circumference of a circle
print(" For a circle with radius {}:".format(radius))    #Output results
print(" The area is:{:.2f}".format(area))
print(" The circumference is:{:.2f}".format(circumference))
```

3-1

在这个程序中,首先导入了Python 的 math 模块,它包含各种数学函数和常数,包括 π。然后,定义了圆的半径(在这个例子中半径为 5),接着使用顺序结构按照公式计算圆的面积和周长,并将结果分别存储在变量 area 和 circumference 中。最后,使用 print()函数输出结果,并保留两位小数。

In this program, first, we import the math module from Python, which includes various mathematical functions and constants, including π. Then, we define the radius of the circle(in this example, it's 5). Next, using sequential structure, we calculate the area and circumference of the circle according to the formulas and store the results in the variables area and circumference, respectively. Finally, we use the print() function to output the results, with two decimal places retained.

3.2 赋值语句
3.2 Assignment Statement

在 Python 中,赋值语句用于将值(可以是变量、表达式或常量)分配给变量。赋值语句使用等号(=)来表示,赋值符号结合性为从右到左,这意味着等号右侧的值或表达式会首先被计算或求值,然后将得到的结果赋值给左侧的变量。以下是 Python 中赋值语句的一些基本介绍和示例。

(1)基本赋值:将单个值赋值给一个变量。在这个例子中,整数 10 被赋值给变量 x。

In Python, an assignment statement is used to assign a value(which can be a variable, an expression, or a constant) to a variable. The assignment statement uses the equal sign (=) to indicate assignment, and the assignment operator has a right-to-left associativity, which means that the value or expression on the right side is evaluated first, and then the result is assigned to the variable on the left side. Here are some basic explanations and examples of assignment statements in Python.

(1)Basic assignment: Assign a single value to a variable. In this example, the integer 10 is assigned to the variable x.

```
x=10
```

(2)多变量同时赋值:Python 允许在一行中同时给多个变量赋值。在本例中,变量 a、b 和 c 分

(2)Multiple variable assignment: Python allows multiple variables to be assigned values in a single line. In this example, variables a, b, and c are assigned values 1,

别被赋值为 1、2 和 3。

2, and 3 respectively.

```
a, b, c=1, 2, 3
```

（3）序列解包：当赋值符号右侧是一个序列（如字符串、列表或元组）时，可以将其解包并赋值给多个变量。在这个例子中，变量 a 被赋值为 p，b 被赋值为 y，c 被赋值为 t。

(3) Sequence unpacking: When the assignment symbol is followed by a sequence(such as a string, list or tuple), it can be unpacked and assigned to multiple variables. In this example, variable a is assigned the value p, b is assigned the value y, and c is assigned the value t.

```
a,b,c="pyt"
print(a,b,c)   #The result is: p y t
```

（4）复合赋值：Python 的复合赋值运算符，如+=、-=、*=、/=等，用于将变量的当前值与赋值运算符右边值进行运算，并将结果重新赋值给复合赋值运算符左侧的变量。在本例中，count 的原始值是 5，经过+=1 操作后，它的值变为 6。

(4) Compound assignment: Python's compound assignment operators, such as +=, -=, *=, /=, etc, are used to perform an operation between the current value of a variable and the value on the right side of the signment operator, and then assign the result back to the variable on the left side of the compound operator. For example, with an original value of 5 for count, after the +=1 operation, its value becomes 6.

```
count=5
count+=1   #Equivalent to count=count+1
```

（5）链式赋值：可以将一个值赋值给多个变量，这些变量将共享相同的值。在本例中，a、b 和 c 都被赋值为 0。

(5)Chained assignment: It allows assigning a single value to multiple variables, where these variables will share the same value. In this example, a, b, and c are all assigned the value 0.

```
a=b=c=0
```

（6）交换变量值：不使用临时变量即可交换两个变量的值。在这个例子中，a 和 b 的值被交换。

(6)Variable swapping: It is possible to exchange the values of two variables without using a temporary variable. In this example, the values of a and b are swapped.

```
a, b=b, a
```

（7）赋值时的类型转换：在赋值过程中，Python 会自动进行某些数据类型的转换。在这个例子中，浮点数 10.5 被转换成了整数 10。

(7)Type conversion during assignment: Python automatically performs certain type conversions during the assignment process. In this example, the floating-point number 10.5 is converted to the integer 10.

```
float_num = 10.5
int_num = int(float_num)    #Explicit type conversion.
```

赋值语句是 Python 中最基本的语句之一，在构建程序逻辑和数据处理时起着至关重要的作用。

Assignment statement is one of the fundamental statements in Python, it plays a crucial role in building program logic and data processing.

3.3 分支结构
3.3 Conditional Structure

Python 的分支结构是控制结构的重要组成部分，它允许程序根据条件的不同执行不同的代码块。其中，单分支结构是最基本的形式，通过 if 语句实现，当条件满足时执行相应的代码。双分支结构则通过 if-else 语句实现，它提供了两种可能的执行路径，根据条件的真假选择其中之一执行。多分支结构则更为复杂，使用 if-elif-else 语句，可以在多个条件中进行选择，并执行对应的代码块。

The conditional structure in Python is an important component of control structures, which allows a program to execute different code blocks based on different conditions. Among them, the single branch structure is the most basic form, implemented using the if statement, which executes the corresponding code when the condition is satisfied. The double branch structure is implemented using the if-else statement, providing two possible execution paths, choosing one to execute based on the truth or falsehood of the condition. The multiple branch structure is more complex, using the if-elif-else statement, which can choose among multiple conditions and execute the corresponding code block.

3.3.1 单分支结构
3.3.1 Single Branch Structure

在 Python 中，单分支结构通常指的是 if 语句，它允许程序在条件为真时，执行一段代码；在条件为假时，退出分支结构。

Python 的单分支结构根据条件判断选择执行或者不执行包含在 if 下方的缩进语句块。这种结构是通过 if 语句来实现的，其基本语法如下：

In Python, a single branch structure typically refers to the if statement, which allows the program to execute a block of code when a condition is true. If the condition is false, the program exits the conditional structure.

The single branch structure in Python is a type of conditional structure that selects whether to execute or not execute the indented statement block beneath the "if" based on a condition. This structure is implemented using the "if" statement, and its basic syntax is as follows:

```
if conditional expression:
    code block            #indent 4 spaces from the letter 'i'
```

下面是一个简单的 Python 单

Here is an example of a simple Python single branch

第3章 控制结构
Chapter 3　Control Structure

分支结构(if 语句)的例子： | structure(if statement):

```
age=18
if age>=18:
    print("You are an adult.")
```

单分支结构通常指的是只有一个 if 条件的控制流结构，如图 3-2 所示。在这个例子中，有一个变量 age，判断它是否大于或等于 18。如果是，则条件为真，执行 if 语句后面的代码块(在这个例子中是 print("You are an adult."))。单分支结构只包含一个条件和一个代码块，如果条件为真，代码块就会被执行；如果条件为假，代码块就会被跳过。

注意，Python 使用缩进来表示代码的从属关系。在这个例子中，print 语句是 if 语句的一部分，因为它被缩进了。

在更复杂的问题中，可能还会遇到双分支结构(if-else 语句)和多分支结构(if-elif-else 语句)，它们允许程序根据多个条件执行不同的代码块。

A single branch structure typically refers to a control flow structure with only one if condition, the execution flow is shown in Figure 3-2. In this example, a variable called age will be checked if it is greater than or equal to 18. If it is, the condition is true, and the code block after the if statement is executed(In this example, it is "print("You are an adult.")"). A single branch structure consists of one condition and one code block, if the condition is true, the code block will be executed; if the condition is false, the code block will be skipped.

The use of indentation in Python indicates the relationship of code. In this example, the print statement is part of the if statement because it is indented.

In more complex problems, one may encounter double branches structures(if-else statements) and multiple branches structures(if-elif-else statements), which allow the program to execute different code blocks based on multiple conditions.

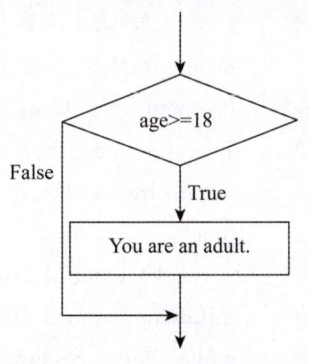

图 3-2　单分支结构流程图

Figure 3-2　Single Branch Structure Flowchart

3.3.2 缩进
3.3.2 Indentation

在 Python 中，缩进是一种极 | Indentation is an extremely important syntax element

其重要的语法元素，它用来表示代码块之间的层次结构。Python 使用缩进来区分代码块的边界，这是它与其他许多语言（如 C++、Java 等）使用大括号{ }来定义代码块的边界的主要区别。缩进规则如下。

（1）一致性：在同一代码块中，所有的语句必须保持相同的缩进级别，否则就会导致"unexpected indent"错误。下面例子就会产生出错提示"unexpected indent"。

in Python. It is used to represent the hierarchical structure between code blocks. In Python, indentation is used to differentiate the boundaries of code blocks, which is a major difference compared to many other languages such as C++ and Java that use curly braces {} to define the boundaries of code blocks. The indentation rules are as follows.

（1）Consistency: In the same code block, all statements must maintain the same level of indentation. Otherwise, it will result in a "unexpected indent" error. The following example will produce an error message "unexpected indent".

```
a=10
if a>0:
    print("{}is a positive number".format(a))
        print("END")
```

（2）通常约定：一般推荐使用 4 个空格作为标准的缩进宽度。虽然 Python 本身允许选择其他的缩进宽度（甚至是混合使用空格和制表符），但这样做可能会导致代码在不同的编辑器或环境中显示缩进不一致，从而产生错误或混淆。

（3）嵌套代码块：当一个代码块嵌套在另一个代码块内部时，内部代码块应该比外部代码块多缩进一个级别。所以内部代码块的每一行都应该比外部代码块多 4 个空格。

（4）控制流语句：if、for、while 等控制流语句后面的代码块需要通过缩进来表示。逻辑上属于同一个执行单元的代码块，缩进级别应该相同。

(2)Conventional rule: It is generally recommended to use 4 spaces as the standard indentation width. Although the Python itself allows for choosing other indentation widths(even a mixture of spaces and tabs), doing so may result in inconsistent indentation display in different editors or environments, leading to errors or confusion.

(3)Nested code blocks: When a code block is nested inside another code block, the inner code block should be indented one level more than the outer code block. Therefore, each line of the inner code block should be indented by additional 4 spaces compared to the outer code block.

（4）Control structure syntax: The code block following control flow statements such as if, for, and while should be indicated through indentation. The code blocks belong to the same execution unit logically, their indentation level should be the same.

3.3.3 双分支结构
3.3.3 Double Branch Structure

Python 的双分支结构是一种根据条件判断选择执行不同代码块的

The double branch structure in Python is a program structure that selects different code blocks to execute

第3章 控制结构
Chapter 3 Control Structure

程序结构，它包含两个可能的执行路径。这种结构是通过 if-else 语句来实现的，其基本语法如下：

based on a condition. It consists of two possible execution paths. This structure is implemented using the if-else statement, and its basic syntax is as follows:

```
if conditional expression:
    code block 1
else:
    code block 2
```

执行流程为：如果条件表达式成立（即结果为 True），则执行 if 下方代码块 1，否则执行 else 下方代码块 2。根据 if 后面条件表达式的真假，程序会选择执行不同的代码块。

下面使用双分支结构实现判断奇偶问题。

【例 3-2】编写一个程序，用户输入一个数字，判断这个数字是奇数还是偶数，并输出相应的信息。

Execution flow is as follows: If the conditional expression evaluates to True, then code block 1 will be executed. Otherwise, code block 2 under the else statement will be executed. Depending on the value of the conditional expression following the if statement, the program will choose to execute different code blocks.

By using a double branch structure, we can solve the problem of checking whether a number is even or odd.

[Example 3-2] Write a program that prompts the user to enter a number, checks whether the number is even or odd, and outputs the corresponding information.

```
#The user inputs a number
num=int(input("Please enter an integer: "))
#Check whether a number is odd or even
if num % 2==0:
    print("{} is an even number.".format(num))
else:
    print("{} is an odd number.".format(num))
```

3-2

代码解释：首先，使用 input() 函数提示用户输入一个整数，并将输入的内容转换为整数类型，存储在变量 num 中。

接着，使用 if-else 语句来判断 num 是奇数还是偶数。这里使用取模运算符%，如果 num 除以 2 的余数为 0，则 if 条件判断为真，说明 num 是偶数，执行 if 下方缩进的代码块中的内容；否则，if 条件为假，执行 else 下方缩进内容。

条件为真，打印 if 下方缩进

Code explanation: First, use the input() function to prompt the user to input an integer, and then convert the input into integer type. Then, store the input into the variable num.

Next, use an if-else statement to check whether the num is odd or even. Here, the modulo operator % is used. If the remainder of num divided by 2 is 0, the if condition is True, indicating that num is an even number, and the code block indented under the if statement will be executed. Otherwise, if the if condition is False, the code block indented under the else statement will be executed.

If the condition is True, print the indented statement

83

语句"{} is an even number."，否则，打印 else 下方的缩进语句"{} is an odd number."。其中的"{}"是一个占位符，它会被后面 print() 函数中 num 变量的值所替换。双分支结构流程图如图 3-3 所示。

"{} is an even number.", Otherwise, print the indented statement "{} is an odd number." The "{}" is a placeholder that will be replaced by the value of the num variable in the print() function. The double branch structure flowchart is shown in Figure 3-3.

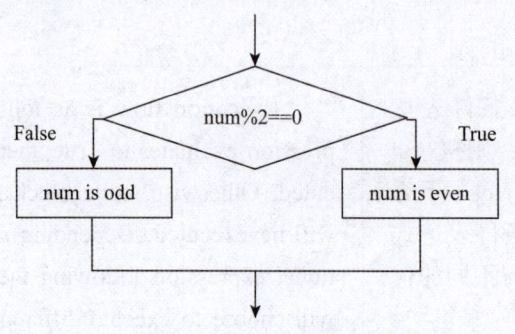

图 3-3 双分支结构流程图

Figure 3-3 Double Branch Structure Flowchart

【例 3-3】使用海伦公式求解三角形面积。已知三角形三边长度 a、b、c，如果 3 条边可以构成三角形，则输出三角形面积，否则，输出"Data Error!"。

海伦公式如下：

[Example 3-3] Calculate the area of a triangle using the Heron's formula. Given the lengths of the three sides a, b, and c of the triangle, if the sides can form a triangle, output the area of the triangle; otherwise, output "Data Error!".

Heron's formula is as follows:

$$s = \sqrt[2]{p*(p-a)*(p-b)*(p-c)}$$

其中，s 是三角形的面积；a、b 和 c 分别是三角形的 3 条边长度；p 是半周长，即 (a+b+c)/2。

计算三角形面积的程序，参考代码如下：

In the formula, s represents the area of the triangle; while a, b, and c are the lengths of the three sides of the triangle; p is the semi-perimeter, which is equal to (a+b+c)/2.

The following code is a reference for calculating the area of a triangle:

```
a,b,c=eval(input("Please enter the three sides of the triangle, separated by commas: "))
p=(a+b+c)/2
if a+b>c and a+c>b and b+c>a:
    s=(p*(p-a)*(p-b)*(p-c))**(1/2)
    print("The area of the triangle is {:.2f}".format(s))
else:
    print("Data Error!")
```

3.3.4 多分支结构
3.3.4 Multi-branch Structure

在 Python 中，多分支结构通过关键字 if、elif 和 else 实现，当用户输入数据后，满足某一个条件则执行相应语句块。例如，根据用户输入的年龄，程序可以判断用户是儿童、青少年还是成年人，并据此执行不同的操作。

当百分制分数按照需求转化为五级分制（A、B、C、D、E）时，根据条件的不同，输出的信息也不同。总之，多分支结构在 Python 中应用范围广，几乎可以应用于任何需要根据不同条件执行不同操作的场景。多分支的语法结构如下：

In Python, the multi-branch structure is implemented using the keywords if, elif, and else. When a user enters data, the program checks for a specific condition and executes the corresponding block of statements. For example, based on the user's input age, the program can determine if the user is a child, teenager, or adult and perform different operations accordingly.

When converting a percentage score in the hundred-point system to a five-level grading system(A, B, C, D, E), the output message varies based on different conditions. In summary, the multi-branch structure is widely used in Python to handle various scenarios where different operations need to be executed based on different conditions. The syntax structure of the multi-branch is as follows:

```
if conditional expression 1:
    code block 1
elif conditional expression 2:
    code block 2
elif conditional expression 3:
    code block 3
…
elif conditional expression n-1:
    code block n-1
else:
    code block n
```

elif 语句和 else 语句是可选的。如果同时省略 elif 语句和 else 语句，只使用 if 语句，此时就变成单分支结构，也可以根据需要添加任意数量的 elif 语句。但是，else 语句应该是多分支结构的最后一个部分，它不需要任何条件表达式，因为它会在前面的所有条件都不满足时执行，else 语句也可根据情况省略。

The elif statements and else statements are optional. If both elif statements and else statements are omitted and only the if statement is used, it becomes a single branch structure. One can add any number of elif statements as needed. However, the else statement should be the last block in a multi-branch structure and does not require any conditional expression, as it will be executed when none of the preceding conditions are satisfied. The else statement can also be omitted as needed.

此外，一旦满足某个条件并执行了相应的代码块，程序就会跳过后续的所有 elif 和 else 代码块，继续执行后面的代码（如果有的话）。这意味着在多分支结构中，只有一个代码块会被执行。多分支结构流程图如图 3-4 所示。

Additionally, once a condition is satisfied and the corresponding code block is executed, the program will skip the subsequent elif and else code blocks and continue executing the following code(if there is any). This means that in a multi-branch structure, only one code block will be executed. The multi-branch structure flowchart is shown in Figure 3-4.

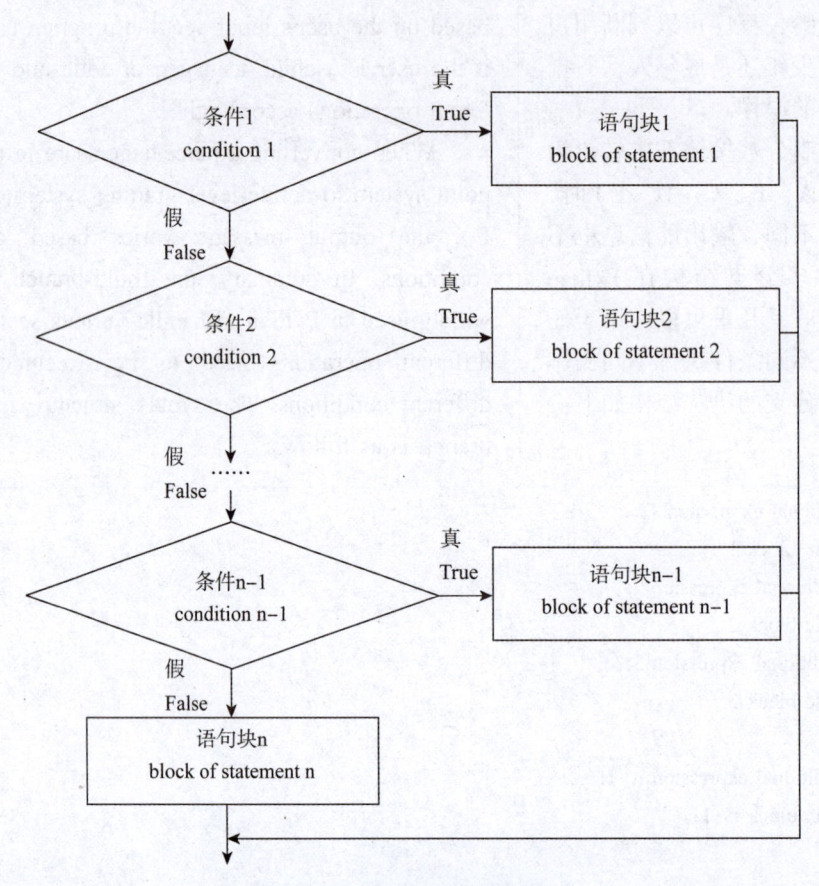

图 3-4 多分支结构流程图

Figure 3-4　Multi-branch Structure Flowchart

【例 3-4】用户输入一个分数（0~100），根据输入的分数判断其等级，并打印出相应的结果。如果分数大于或等于 90，则输出"A"。如果分数在 80 到 90 之间，则输出"B"。如果分数在 70 到 80 之间，则输出"C"。如果分数在 60 到 70 之间，则输出"D"。如果分数低于 60，则输出"E"。

[Example 3-4] The user enters a score(0-100), and based on the entered score, determine its level and print the corresponding result. If the score is greater than or equal to 90, output "A". If the score is between 80 and 90, output "B". If the score is between 70 and 80, output "C". If the score is between 60 and 70, output "D". If the score is below 60, output "E".

第3章　控制结构
Chapter 3　Control Structure

```
score=eval(input("Please enter a score(0~100): "))
if score>=90:
    print("A")
elif score>=80:
    print("B")
elif score>=70:
    print("C")
elif score>=60:
    print("D")
else:
    print("E")
```

3-4

【例3-5】分段函数如下，根据输入的 x 值返回不同的 y 值，通过多分支结构来实现问题的求解。

[Example 3-5] A piecewise function is as fllows, which returns different y-value based on the input x-value, and the problem can be solved using a multi-branch approach.

$$y = \begin{cases} -3x+1, & x<-2 \\ x^2, & -2 \leqslant x<3 \\ 2x-5, & x \geqslant 3 \end{cases}$$

```
#Accept the user input for the value of x
x=eval(input("Please enter the value of x:"))
#Calculate the value of y using a multi-branch structure
if x<-2:
    y=-3*x+1
elif -2<=x<3:
    y=x**2
else:   #x>=3
    y=2*x-5
#Output the value of y
print("y={}".format(y))
```

3-5

3.3.5　分支结构嵌套
3.3.5　Nested Conditional Structure

在 Python 中，分支结构嵌套指的是在一个条件语句的内部包含另一个或多个分支结构。分支结构嵌套可以根据多个条件进行更复杂的决策。分支结构嵌套多种多样，下面将通过例题介绍分支结构嵌套

In Python, nested conditional structure refers to including another or multiple branch structures within a conditional statement. Nested conditional structures allow for more complex decisions based on multiple conditions. There are various ways to nasted conditional structures, the specific usage of nested conditional structures will be

的具体使用。

【例 3-6】编写一个程序，该程序接收用户输入的 3 个整数代表三角形三边长，并根据以下规则判断它们是否能构成一个三角形，如果可以构成三角形，则判断其是等边三角形或等腰三角形或不等边三角形。如果不能构成三角形，则输出"Data ERROR！"。

[Example 3-6] Write a program that takes three integers as input from the user to represent the lengths of the three sides of a triangle. Based on the following rules, check whether they can form a triangle. If a triangle can be formed, check whether it is an equilateral triangle, an isosceles triangle, or a scalene triangle. If a triangle cannot be formed, output "Data ERROR！".

```
#Receive three integers inputted by the user
a=int(input("Please enter the first integer:"))
b=int(input("Please enter the second integer:"))
c=int(input("Please enter the third integer:"))
#Determine if they can form a triangle
if (a+b>c) and (a+c>b) and (b+c>a):
    print("The input sides can form a triangle. ")
    #Determine the type of the triangle
    if a==b==c:
        print("This is an equilateral triangle. ")
    elif (a==b and a !=c) or (a==c and a !=b) or (b==c and b !=a):
        print("This is an isosceles triangle. ")
    else:
        print("This is a scalene triangle. ")
else:
    print("Data ERROR!")
```

3-6

常见的分支结构嵌套语法如下：

Common nested branch structure syntax is as follows:

```
if conditional statement 1:
    if conditional statement 2:
        block of code 1
    else:
        block of code 2
```

```
if conditional statement 1:
    if conditional statement 2:
        block of code 1
    else:
        block of code 2
else:
```

```
        if conditional statement 3:
            block of code 3
        else:
            block of code 4
```

用户可以根据需要嵌套任意多的分支结构，但过多的嵌套可能会使代码难以理解和维护。嵌套越多，代码的缩进量越多，一旦缩进出错，很难找到问题所在。

Users can nest as many branch structures as needed, but excessive nesting can make the code difficult to understand and maintain. The more nested the structures, the more the code indentation increases, making it challenging to identify the source of errors if indentation is incorrect.

3.4 循环结构
3.4 Loop Structure

循环结构是编程中的一种基本控制结构，用于重复执行一段代码，直到满足特定条件为止。通过设置循环条件和循环体，可以实现某段代码的重复执行，提高程序的执行效率。常见的循环结构包括 for 循环和 while 循环。for 循环主要用于循环次数确定的重复执行，而 while 循环则根据条件控制反复执行代码块。

The loop structure is a fundamental control structure in programming, used to repeat a code block until a specific condition is met. By setting the loop condition and loop body, it enables the repetition of a code block and improves the efficiency of program execution. Common loop structures include the for loop and the while loop. The for loop is primarily used for repeating a code block with a known number of iterations, the while loop repeatedly executes a code block based on a condition.

3.4.1 range() 函数
3.4.1 range() Function

在 Python 中，for 循环和 range() 函数经常一起使用，range() 函数用于生成一个数字序列，而 for 循环则遍历这个序列中的每一个元素。

range() 函数用于生成一个不可变的数值序列。range() 函数提供 3 个参数：start、stop 和 step，只有 stop 参数不可以省略。

start（可选）：序列的起始值，默认为 0。stop（不可省略）：序列的终止值，但 range() 产生的终值不

In Python, the for loop and the range() function are often used together. The range() function generates a sequence of numbers, and the for loop iterates over each element in this sequence.

The range() function is a built-in function used to generate an immutable sequence of numerical values. The range() function has three parameters: start, stop, and step, only the stop parameter is mandatory and cannot be omitted.

start(optional): The start value of the sequence, defaults to 0. stop(required): The end value of the sequence, but the value itself is not included in the range()

包含此值本身。step（可选）：序列中每个数字之间的差，即步长，默认为 1。

range()函数返回一个"range 对象"，它是一个不可变的惰性序列，这意味着它只表示值的范围，而不实际存储这些值。range()函数的语法如下：

generated sequence. step (optional): The difference between each number in the sequence, defaults to 1.

The range() function returns a "range object", which is an immutable and lazy sequence. This means that it only represents the range of values without actually storing these values. The syntax of the range() function is as follows:

> range(start,stop,step)
> #Generate an arithmetic sequence with values ranging from 'start' to 'stop-1', with a step size of 'step'.

start、stop 和 step 这 3 个参数必须都为整数。其中，step 可以是正整数或者负整数，但是不可以为 0。

The three parameters, start, stop, and step, must all be integers. The step can be a positive or negative integer, but it cannot be zero.

> range(0,5,1) #If the start is 0 and the step is 1, both the start and step parameters can be omitted, which is
> #equivalent to range(5).

range(5)将产生 0 到 4 步长为 1 的 5 个元素，不包括结束参数 5，通过 print()函数可以将这 5 个元素打印输出，但是要在 range(5)前面加一个"＊"号。"＊"号用作解包操作符，它可以将一个可迭代对象（如 range()函数、列表、元组、字符串或任何迭代器）中的元素"解包"并传递给函数。当使用 print(＊range(5))时，实际上是将 range(5)生成的每个元素作为独立的参数传递给输出函数，最后输出。

打印输出 range(5)元素如图 3-5 所示。

The range(5) function generates 5 elements from 0 to 4 with a step size of 1, excluding the end parameter 5. By using the print() function, these 5 elements can be printed out. However, one needs to add an asterisk(*) before range(5). The asterisk(*) serves as an unpacking operator, which can "unpack" the elements from an iterable object(such as the range() function, lists, tuples, strings, or any iterator) and pass them as separate parameters to a function. When using print(*range(5)), each element generated by range(5) is actually passed as an individual argument to the print function, resulting in the desired output.

Print the elements of range(5) is shown in Figure 3-5.

```
>>> print(*range(5))
0 1 2 3 4
```

图 3-5 打印输出 range(5)元素
Figure 3-5 Print the Elements of range(5)

因为 print()函数可以接收多个

Because the print() function can accept multiple ar-

参数，并将它们打印在同一行上，所以输出是：0 1 2 3 4。

每个数字都是由 print() 函数以空格分隔打印出来的。这个操作等效于手动将范围中的每个数字作为独立参数传递给 print()，如 print(0，1，2，3，4)。解包操作符"＊"，提供了一种简洁的方式来处理可迭代对象中的元素。

当开始参数不为默认值 0 时，不可省略。步长不为默认值 1 时，步长也同样不能省略。

guments and print them on the same line, the output is: 0 1 2 3 4.

Each number is printed by the print() function separated by a space. This operation is equivalent to manually passing each number in the range as separate arguments to the print() function, like print(0, 1, 2, 3, 4). The unpacking operator "*", provides a concise way to handle elements in an iterable object.

When the start parameter is not the default value 0, it cannot be omitted. Similarly, when the step parameter is not the default value 1, the step cannot be omitted.

```
range(1,10,2)    #1,3,5,7,9
range(1,10,3)    #1,4,7
```

注意，range() 函数中的第 3 个参数步长可以为负数，当步长参数为负时，设置的开始参数要比结束参数值大，此时，生成的元素按降序排列。

Note that the third parameter, step, in the range() function can be a negative number. When the step parameter is negative, the start parameter should be greater than the end parameter, and in this case, the generated elements will be in descending order.

```
range(10,1,-2)     #10 8 6 4 2
range(10,1,-3)     #10 7 4
range(-1,-10,-1)   #-1 -2 -3 -4 -5 -6 -7 -8 -9
```

如果步长设置为 0，则会出现"ValueError"的错误提示，如图 3-6 所示。

If the step is set to 0, it will raise a "ValueError" error. The error message shows in Figure 3-6.

```
>>> print(*range(1,5,0))
Traceback (most recent call last):
  File "<pyshell#0>", line 1, in <module>
    print(*range(1,5,0))
ValueError: range() arg 3 must not be zero
```

图 3-6　步长为 0 的报错提示

Figure 3-6　The Error Message for a Step of 0

3.4.2　for 循环
3.4.2　for Loop

在 Python 中，for 循环是一种基本的控制流结构，用于遍历序列，如 range() 函数、列表、字符串等。

In Python, a for loop is a basic control flow structure used for iterating over sequences such as range() function, lists, strings, etc.

在 for 循环中，循环控制变量按顺序访问可迭代对象中的每个元素，并对每个元素执行相同的操作。for 循环基本语法如下：

In a for loop, the loop control variable sequentially accesses each element in an iterable object and performs the same operation on each element. The basic syntax of a for loop is as follows:

```
for variable in iterable:
    loop body    #indented loop body
```

在 for 循环语法结构中，最重要的两个关键字是 for 和 in，以 for 开始的第一行代码称为循环的开始，冒号下方的循环体语句需要相对于 for 关键字的字母 f 向右缩进 4 个英文空格，这与 if 结构类似。并且包含在同一循环体的语句缩进量相同。

The two most important keywords in the syntax structure of a for loop are "for" and "in". The first line of code that starts with "for" is called the start of the loop. The statements in the loop body, which are located below the colon, need to be indented four spaces to the right relative to the letter "f" in the "for" keyword, similar to the indentation for "if" structure. Furthermore, statements that belong to the same loop have the same level of indentation.

在第一行代码中，for 循环中的 variable 是一个循环控制变量，用于在每次迭代中存储当前元素的值。iterable 是指循环控制变量要遍历的 range() 函数、字符串、列表、元组、字典、集合、文件。

The variable in the first line of code is a loop control variable, which is used to store the value of the current element in each iteration. The iterable refers to the range() function, string, list, tuple, dictionary, set, or file that the loop control variable is to iterate over.

循环控制变量命名尽量简单，通常使用单个字母，例如 i 或者 j 等。循环体语句可以是一条也可以是多条，多条语句缩进量相同。例如：

The loop control variable is usually named using a single letter, such as "i" or "j". The body of the loop can contain either a single statement or multiple statements, with the same indentation level. For example:

```
for i in range(5):      #i is the loop control variable
    print(i)            #print(i) is the loop body statement, indented 4 spaces to the right of "f"
```

变量 i 每次进入循环前从 range(5) 中取值，第一次进入循环之前，i 值为 0，循环体执行打印操作，输出 0。第二次进入循环体时，i 从 range(5) 中取第 2 个元素，此时 i 值为 1，再次执行循环体语句，打印输出 i 值，即输出结果 1。

The variable "i" takes values from the range(5) before each iteration. Before the first iteration, the value of "i" is 0, and the loop body executes a print operation, outputting 0. In the second iteration, "i" takes the second element from the range(5), which is 1, and the loop body executes the print statement again, outputting the result of 1.

以此类推，重复执行 print(i) 操作，输出 5 个值，依次为 0 ~ 4，注意，每打印一个数值，执行换行

This process repeats, printing five values from 0 to 4. It's important to note that after printing each value, a line break is executed because the print() function, when

第3章 控制结构
Chapter 3　Control Structure

操作，其原因是 print() 函数的 end 参数省略时，默认结束符为换行符。循环执行流程如图 3-7 所示。

the "end" parameter is omitted, defaults to ending with a line break. The process of loop execution shown in Figure 3-7.

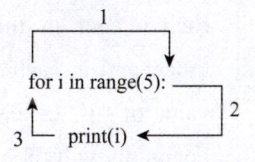

```
       1
    for i in range(5):
                        2
 3 ← print(i) ←
```

图 3-7　循环执行流程

Figure 3-7　The Process of Loop Execution

for 循环的执行流程：首先判断可迭代序列中是否有元素，如果有，则获取可迭代对象中的第 1 个元素并将其赋值给循环控制变量；然后执行循环体内的代码；接着再次从可迭代对象中获取下一个元素，直到所有元素都被遍历完；最后结束循环，执行 for 循环外的代码。for 循环执行流程如图 3-8 所示。

The execution process of a for loop is as follows: First, it checks if there is any element that can be obtained from the iterable sequence. If there is, it assigns the first element to the loop control variable. Then, it executes the code within the loop body. Next, it fetches the next element from the iterable object until all elements have been traversed. Finally, it exits the loop and executes the code outside the for loop. The process of executing a for loop is shown in Figure 3-8.

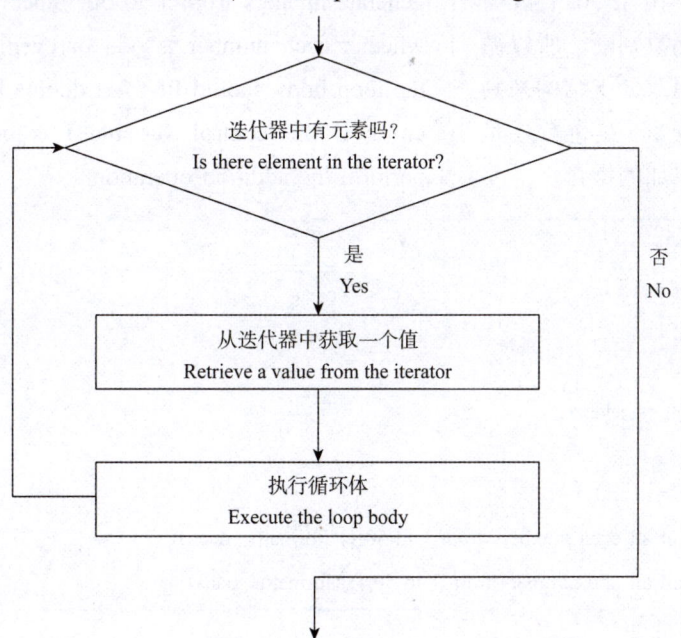

图 3-8　for 循环执行流程

Figure 3-8　The Process of Executing a for Loop

【例 3-7】使用 for 循环求 1~50 中所有整数的和。

[Example 3-7] Use a for loop to find the sum of all integers from 1 to 50.

93

算法分析：首先使用range(1, 51)函数构造1~50个整数，循环控制变量i依次获取迭代器中的每一个整数，循环执行50次，每次循环要做的是不断向一个求和变量s中累加当前循环控制变量i的值，执行加法操作50次。循环结束后，打印输出s的值。

Algorithm analysis: First, use the range (1, 51) function to generate a sequence of integers from 1 to 50. Then, use a loop control variable "i" to iterate through each integer in the iterator. The loop will execute 50 times, and in each iteration, continuously add the current value of "i" to a summation variable "s", this addition operation will be performed 50 times. After the loop ends, print the value of "s".

```
s=0
for i in range(1,51):
    s=s+i
print(s)
```

3-7

【例3-8】使用for循环分别求1~50中所有奇数和偶数的和。

[Example 3-8] Use a for loop to find the sum of all odd numbers and even numbers separately from 1 to 50.

算法分析：使用range(1,51)构造1~50个整数，由于50个数需要进行奇数或者偶数判断，所以循环体语句需要使用双分支结构来判断当前循环控制变量i是奇数还是偶数，再去执行累加的操作。

Algorithm analysis: Use the range(1,51) function to generate integers from 1 to 50. Since we need to check whether each number is odd or even, the statements in the loop body should first use double branch structure to check if the control variable i is odd or even, then perform the addition operation.

```
s_odd=0
s_even=0
for i in range(1,51):
    if i%2==0:
        s_even=s_even+i
    else:
        s_odd=s_odd+i
print("The sum of all even numbers from 1 to 50:{}".format(s_even))
print("The sum of all odd numbers from 1 to 50:{}".format(s_odd))
```

3-8

【例3-9】构造斐波那契数列前10项：1，1，2，3，5，8…，斐波那契数列前2项已知，均为1，从第3项开始每项都是前2项之和，以此类推。

[Example 3-9] Construct the Fibonacci sequence with the first 10 terms: 1, 1, 2, 3, 5, 8…, where the first two terms of the Fibonacci sequence are known to be 1, and each subsequent term is the sum of the two preceding terms.

第3章 控制结构
Chapter 3 Control Structure

算法分析：定义2个变量a和b，分别作为斐波那契数列的前两项1和1。直接打印出a和b的值，作为数列的开始。

迭代计算：使用for循环，从第3个数开始迭代（因为前两个数已经打印），直到生成所需的数列长度（在这个例子中是10项，但由于已经打印了2项，所以循环8次）。

更新变量：在每次循环中，根据斐波那契数列的定义 $F(n)=F(n-1)+F(n-2)$，更新a和b的值。这里使用元组解包的方式同时更新a和b，a被赋值为b的当前值，b被赋值为a、b之和。

打印结果：在每次循环中，打印出新的斐波那契数列新项（存储在变量b中）。

Algorithm analysis: Define two variables a and b, representing the first two terms of the Fibonacci sequence as 1 and 1. Print the values of a and b directly as the starting numbers of the sequence.

Iterative calculation: Using a for loop, iterate from the third number onwards (since the first two numbers have already been printed), until the desired length of the sequence is generated(In this example, 10 terms, but since 2 terms have already been printed, the loop will iterate 8 times).

Updating variables: In each iteration, update the values of a and b based on the definition of the Fibonacci sequence $F(n)=F(n-1)+F(n-2)$. Here, we use the method of tuple unpacking to update both a and b simultaneously, a is assigned the current value of b, while b is assigned the sum of a and b.

Printing the result: In each iteration, print the new Fibonacci sequence term(stored in the variable b).

```
#Initializing the first two terms of the Fibonacci sequence
a, b = 1, 1
#Printing the first two terms
print(a, end=',')
print(b, end=',')
#Using a for loop to calculate and print the subsequent terms
for i in range(2, 10):    #Generating the next 8 terms of the Fibonacci sequence
    #Calculating the next Fibonacci term
    a, b = b, a+b
    #Printing the current Fibonacci term
    print(b, end=',')
#The output result is: 1,1,2,3,5,8,13,21,34,55,
```

3-9

3.4.3 while 循环
3.4.3 while Loop

while 循环首先根据逻辑表达式构造条件判断，决定是否执行循环体语句，直到循环条件为假，结束 while 循环。while 循环通常

The while loop first constructs a condition based on a logical expression to determine whether to execute the statements within the loop. It continues to execute the loop as long as the condition evaluates to true, and stops

95

用于循环次数不确定，但明确循环应该在满足某个条件时停止的情况。while 循环语法如下：

when the condition becomes false. The while loop is typically used when the number of iterations is uncertain, but it is clear that the loop should stop when a certain condition is met. The syntax of the while loop is as follows:

```
while the loop condition:     #The logical expression
    the loop body             #the code block to be executed
```

在循环次数确定的情况下，可以使用 for 和 while 循环，使用 for 循环解决的问题一般都可以使用 while 循环进行改写。while 循环执行流程如图 3-9 所示。

In situations where the number of iterations is known, both for and while loops can be used. Problems solved using for loops can generally be rewritten using while loops. The execution flow of a while loop is shown in Figure 3-9.

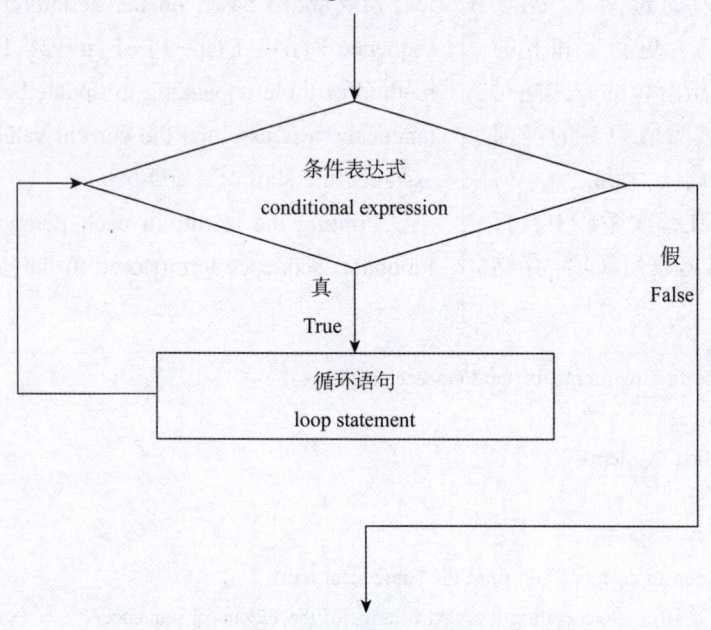

图 3-9　while 循环执行流程

Figure 3-9　The Execution Flow of a while Loop

【例 3-10】使用 while 循环和 for 循环实现打印数字 1 到 5。

[Example 3-10] Print numbers 1 to 5 using while loop and for loop.

```
#Using a while loop
i=1
while i<=5:
    print(i)
    i+=1
#Using a for loop
for i in range(1,6):
    print(i)
```

3-10

while 循环会持续执行直到变量 i 的值超过 5，当"i<=5"条件为假时，结束 while 循环。

while 循环次数可以是 0 次也可以是多次，如果一开始 while 后面的条件为假，则不进入循环。如果 while 循环的条件表达式一直为真，则称 while 循环为无限循环，终止无限循环的方式为〈Ctrl+C〉组合键。

The while loop will continue executing until the value of the variable i exceeds 5. Once the condition "i<=5" becomes false, the while loop will end.

While loop can be executed zero or multiple times. If the condition in the while statement is false, the loop will not be entered. A while loop is called an infinite loop if the condition expression is always true. To terminate an infinite loop, one can use the 〈Ctrl+C〉 keyboard combination.

```
i = 1
while i>0:
    print(i)
    i = i+1
```

while 循环与 for 循环的不同之处如下。

控制方式：while 循环通过条件表达式来控制循环的执行；for 循环则通过遍历可迭代对象的每个元素来控制循环的执行。

循环次数：while 循环的循环次数一般是不确定的，它依赖于条件的真假；for 循环的循环次数通常是确定的，因为它与可迭代对象的长度或元素数量直接相关。

结构：while 循环需要显式地更新循环控制变量以改变循环条件；for 循环则自动处理循环控制变量的更新，直到可迭代对象遍历结束。

【例 3-11】使用 while 循环求 1~100 中所有整数的和。

初始化：在开始循环之前，设置两个变量。total_sum 用于存放所有整数的总和，初始化为 0。i 是循环控制变量，用来控制循环次数，由于是 100 个整数相加（不包

The differences between a while loop and a for loop are as follows.

Control method: While loop controls the execution of the loop through a conditional expression; for loop controls the execution of the loop by iterating through each element of an iterable object.

Number of iterations: The number of iterations in a while loop is generally uncertain, as it depends on the truth or falsity of the condition. The number of iterations in a for loop is typically predetermined, as it is directly related to the length or number of elements in the iterable object.

Structure: In a while loop, the loop control variable needs to be explicitly updated to change the loop condition from true to false. In a for loop, the loop control variable is automatically updated until the iteration of the iterable object is completed.

[Example 3-11] Use a while loop to calculate the sum of all integers from 1 to 100.

Initialization: Before starting the loop, set two variables. total_sum is used to store the total sum of all integers, initialized to 0. i is the loop control variable, which controls the number of loops. Since we are adding 100 integers(excluding 0), the initial value of i is set to 1.

含0），所以i的初始值设置为1。

循环条件：while 循环的条件是i<=100。只要i的值小于或等于100，循环就会继续执行。一旦i的值超过100，循环就会停止。

循环体：在每次循环中，执行两个操作。首先，将i的当前值加到变量total_sum中。然后，递增i的值，以确保在下一次循环中累加更新之后的i值。

终止条件：循环将继续执行，直到i的值变为101，此时i<=100的条件不再满足，条件为假，循环终止。

输出结果：循环结束后，打印输出 total_sum 的值，即1到100之间所有整数的和。

Loop condition: The condition of the while loop is i<=100. As long as the value of i is less than or equal to 100, the loop will continue. Once the value of i exceeds 100, the loop will stop.

Loop body: In each iteration of the loop, two operations are performed. First, the current value of i is added to the variable total_sum. Then, the value of i is incremented to ensure that the updated value of i is accumulated in the next iteration.

Termination condition: The loop will continue to execute until the value of i becomes 101. At this point, the condition i<=100 is no longer satisfied, resulting in a false condition, and the loop terminates.

Output result: After the loop ends, print out the value of total_sum, which is the sum of all integers between 1 and 100.

```python
#Initialize variables
total_sum=0          #To store the sum of integers between 1 and 100
i=1                  #Loop control variable, starting from 1
#while loop
while i<=100:        #The executable condition of the loop is that i is less than or equal to 100
    total_sum+=i     #Add i to the total_sum
    i+=1             #Increment the loop control variable i by 1
#Output result
print("The sum of all integers between 1 and 100: ", total_sum)
```

3-11

【例3-12】求1到100之间能被5整除但不能被7整除的所有数的和。

初始化：在开始循环之前，设置两个变量。total_sum 用于存放所有满足条件的整数的总和，初始化为0。i 是循环控制变量，用来控制循环次数，由于1到100之间的所有整数都需要遍历，所以i初始化为1。

循环条件：while 的循环条件是i<=100。

[Example 3-12] Find the sum of all numbers between 1 and 100 that are divisible by 5 but not divisible by 7.

Initialization: Before starting the loop, set two variables. total_sum is used to store the total sum of all integers that satisfy the condition, initialized as 0. i is the loop control variable, used to control the number of iterations. Because all integers between 1 and 100 need to be traversed, i is initialized as 1.

Loop condition: The loop condition of the while is i<=100.

Loop body: In each iteration, two operations are performed. First, the current value of i entering the loop is used for an if condition. If i is divisible by 5 but not divisible by 7, the value of i is added to the variable total_sum. Then, the value of i is incremented.

Termination condition: When the value of i becomes 101, the condition i<=100 is no longer satisfied, and the condition is false, leading to the termination of the loop.

```
total_sum=0              #To store the sum of integers that meet a certain condition
i=1                      #Loop control variable, starting from 1
#while loop
while i<=100:            #The executable condition of the loop is that i is less than or equal to 100
    if i%5==0 and i%7!=0:
        total_sum +=i    #Add i to the total_sum
    i +=1                #Increment the loop control variable i by 1
print("total_sum=", total_sum)
```

[Example 3-13] Given a polynomial $\frac{\pi}{4} = 1 - \frac{1}{3} + \frac{1}{5} - \frac{1}{7} + \cdots$, find the approximate value of π(pi) until the last term is less than 10^{-6}.

Algorithm analysis: This algorithm can use a variable to track the sign(positive or negative) of the current term, and update the sign and denominator in each iteration. The loop will stop when the absolute value of a term is less than 10^{-6}.

```
pi_approx=0              #Approximation of the value of π
sign=1                   #Since the first term is positive, it is initialized as 1, which is used to
                         #alternate the sign of each term
denominator=1            #It is used to construct the denominator of each term, where the denominator
                         #of the first term starts from 1
term=1.0                 #The initial value of the current term is 1/1
#Using a while loop to calculate an approximation of π/4
while abs(term)>=10e-6:  #Using a while loop to calculate an approximation of π until the absolute
                         #value of term is less than 0.000001
    pi_approx +=sign*term  #Adding or subtracting the current term based on its sign
```

```
#Update the sign and denominator to calculate the next term
sign*=-1                        #Update the sign
denominator +=2                 #Update the denominator
term=1.0/denominator            #Calculate the value of the next term
#Due to the calculation is for the value of π/4, it is necessary to multiply the result by 4
pi_approx*=4
print("The approximate value of π is:",pi_approx)
```

3-13

代码中使用 abs(term)>=10e-6 作为循环条件，以确保在最后一项的绝对值小于 0.000 001 时退出循环。在每次循环中，将当前项加到 pi_approx 上，然后更新符号和分母以准备下一项的计算。最后，将 pi_approx 乘以 4 以得到 π 的近似值。

In the code, abs(term)>=10e-6 is used as the loop condition to ensure that the loop exits when the absolute value of the last term is less than 0.000,001. In each iteration of the loop, the current term is added to pi_approx, and then the sign and denominator are updated to prepare for the calculation of the next term. Finally, pi_approx is multiplied by 4 to obtain an approximate value of π.

3.4.4 break 语句
3.4.4 break Statement

break 语句用于提前终止当前循环的执行，并跳出该循环体。

无论是 for 循环还是 while 循环，当程序遇到 break 语句时，都会立即停止当前循环的执行，并继续执行循环之后的代码。

break 语句通常与分支结构一起使用，在满足特定条件时中断循环。下面是一个使用 break 语句的示例。

The break statement is used to terminate the current execution of a loop and jump out of the loop body.

Whether it is a for loop or a while loop, when the program encounters a break statement, it immediately stops the execution of the current loop and proceeds to execute the code after the loop.

The break statement is commonly used in conjunction with conditional structures to interrupt a loop when certain conditions are met. Here is a examples of using the break statement.

```
#Using break in a for loop
for i in range(10):
    if i==5:
        break
    print(i,end=' ')
```

输出：
0 1 2 3 4

Output:
0 1 2 3 4

在这个例子中，当 i 等于 5 时，条件判断为真，break 语句被执行，

In this example, when i equals 5 and the condition evaluates to true, the break statement is executed, and the

第3章 控制结构
Chapter 3 Control Structure

循环将提前终止，因此只有 0~4 被打印出来。

【例 3-14】判断整数 n 是否是素数，如果是，则输出"n is a prime number"，否则输出"n is not a prime number"。

算法分析：首先，获取输入的整数 n。然后，用一个标志变量 is_prime 来标记 n 是否是素数。接下来，使用一个 for 循环来检查从 2 到 n-1 的所有整数，看它们是否能够整除 n。

如果找到了一个能够整除 n 的整数，则说明该整数不是素数，将 is_prime 设置为 False，并使用 break 语句跳出循环。最后，根据 is_prime 的值（True 或 False）输出相应的结果。

loop is terminated prematurely. Therefore, only numbers from 0 to 4 are printed.

[Example 3-14] Determine whether the integer n is a prime number. If it is, output "n is a prime number"; otherwise, output "n is not a prime number".

Algorithm analysis: First, get the input integer n. Then, use a flag variable is_prime to indicate whether n is a prime number. Next, use a for loop to check all integers from 2 to n-1 to see if they can divide n evenly.

If an integer that can divide n evenly is found, it indicates that the integer is not a prime number, set is_prime to False and break out of the loop using the break statement. Finally, output the corresponding result based on the value of is_prime(True or False).

```
n=int(input("Please enter an integer: "))
if n<=1:
    print("{} is not a prime number ".format(n))
else:
    is_prime=True                #Assume that n is a prime number
    for i in range(2,n):         #It only needs to check the range [2, n-1]
        if n%i==0:               #If n is divisible by i
            is_prime=False       #n is not a prime number
            break                #Break out of the loop
    if is_prime:
        print("{} is a prime number ".format(n))
    else:
        print("{} is not a prime number ".format(n))
```

3-14

break 语句只能跳出最近的一层循环，且只能在循环体内部使用，不能在循环体外部单独使用，否则会引发"SyntaxError"错误。

除了 break 语句，Python 还有 continue 语句，它用于跳过当前循环的剩余语句的执行，提前进入下一次循环。但是 continue 不会终止

The break statement can only jump out of the innermost loop, and the break statement can only be used within a loop body. It cannot be used alone outside a loop body, otherwise it will cause a "SyntaxError" error.

In addition to the break statement, Python also has the continue statement, which is used to skip the execution of the remaining statements in the current iteration of a loop and proceed to the next iteration. However,

101

整个循环的执行。

the continue statement does not terminate the entire loop execution.

3.4.5 continue 语句
3.4.5 continue Statement

continue 语句用于控制循环的执行流程。当循环中满足某个条件时，会触发 continue 语句执行，此时会立即停止当前循环体内 continue 后面剩余语句的执行，并跳转到下一次循环开始的位置。continue 语句通常用于跳过循环中某些不需要执行的代码块。

The continue statement is used to control the execution flow of a loop. When a certain condition is met within the loop, the continue statement is triggered. At this point, the execution of the remaining statements after continue within the current loop iteration is immediately stopped, and the program proceeds to the beginning of the next iteration. The continue statement is often used to skip certain code blocks within a loop that do not need to be executed.

continue 语句在 for 和 while 循环中都可以使用。下面是一个使用 continue 语句的示例。

The continue statement can be used in both for and while loops. Here is an example of using the continue statement.

```
#Print all odd numbers within 10
for i in range(1, 11):
    if i % 2 == 0:        #If i is an even number
        continue           #Skip the remaining part of the current loop
    print(i)              #Print odd numbers
```

在上面的例题中，continue 语句用于跳过所有偶数，因此只有奇数被打印出来。

The above example demonstrates the use of the continue statement to skip all even numbers, resulting in only odd numbers being printed.

continue 语句只能用于循环体内部，不能在循环体外部单独使用，否则会引发"SyntaxError"错误。continue 语句通常与 if 语句结合使用，以根据特定条件跳过循环中的某几次循环。

The continue statement can only be used within the body of a loop and cannot be used outside the loop, as it would cause a "SyntaxError" error. The continue statement is typically used in conjunction with an if statement to skip certain iterations of the loop based on specific conditions.

【例 3-15】输出 40 以内不能被 3 整除的正整数。

算法分析：通过 for 循环遍历 40 以内所有正整数，构造 range(1,41) 可迭代对象。循环内判断当前循环控制变量 i 是否能被 3 整除，如果是，则不输出当前 i 值，使用

[Example 3-15] Output positive integers within 40 that are not divisible by 3.

Algorithm analysis: Iterate through all positive integers within 40 using a for loop by creating an iterable object with range(1, 41). Within the loop, check if the current loop control variable i is divisible by 3. If it is, skip the output of the current i using continue. Otherwise,

continue 来跳过输出。否则，continue 不执行，输出当前 i 值。

continue statement is not executed, output the current i.

```
for num in range(1, 41):
    #Check if the num is divisible by 3
    if num % 3 == 0:
        continue   #If the num is divisible by 3, skip the current number
    #Output the numbers that is not divisible by 3
    print(num, end=' ')
```

3-15

3.4.6 pass 语句
3.4.6 pass Statement

在 Python 中，pass 语句是一个空操作语句，pass 语句的存在是为了保持程序结构的完整性。

pass 语句的用途包括占位符：当正在编写代码，但还没决定好某个部分应该怎么写时，可以暂时用 pass 语句占位。在不影响程序其余部分的情况下继续编写或测试其他代码。有些 Python 结构，如 if 语句、for 循环、while 循环、函数定义、类定义等在语法上要求必须保持结构完整。如果暂时不需要这些语句做任何事情，则可以使用 pass 语句进行占位。

pass 语句可以出现在 Python 代码的任何地方。

In Python, the pass statement is a null operation that serves the purpose of maintaining the integrity of the program structure.

Its usage includes placeholder: When writing code and haven't yet decided how a certain part should be implemented, the pass statement can be temporarily used as a placeholder. This allows one to continue writing or testing other code without affecting the rest of the program. Some Python structures, such as if statements, for loops, while loops, function definitions, class definitions, etc, require maintaining a complete structure in terms of syntax. If there is no need to perform any action for these statements at the moment, the pass statement can be used as a placeholder.

The pass statement can appear anywhere in Python code.

```
for i in range(5):
    pass   #Not yet decided what the loop body should do
```

```
if some_condition:
    pass   #Not yet determined what to do when the condition is met
else:
    print("Condition is false")
```

3.4.7 循环嵌套
3.4.7 Nested Loop

循环嵌套是指在一个循环体内

Nested loops refer to placing one or more loops in-

103

再嵌套另一个或多个循环体，这种结构通常用于处理更复杂的循环逻辑，例如创建复杂的图案等。循环嵌套会增加计算的复杂性，可能导致程序运行变慢，特别是当循环次数很多时。当循环嵌套增多时，层次缩进也会变多，此时程序结构会变得复杂。

【例3-16】打印如图3-10所示的九九乘法表。

side another loop within a loop body, and this structure is typically used to handle more complex looping logic, such as generating intricate patterns. Nested loops can increase the complexity of computations, potentially leading to slower program execution, especially when the number of iterations is high. When the nested loops increase, the level of indentation also grows, making the program structure more complex.

[Example 3-16] Print the multiplication table of 9 as shown in the Figure 3-10.

```
1 × 1 =  1
2 × 1 =  2  2 × 2 =  4
3 × 1 =  3  3 × 2 =  6  3 × 3 =  9
4 × 1 =  4  4 × 2 =  8  4 × 3 = 12  4 × 4 = 16
5 × 1 =  5  5 × 2 = 10  5 × 3 = 15  5 × 4 = 20  5 × 5 = 25
6 × 1 =  6  6 × 2 = 12  6 × 3 = 18  6 × 4 = 24  6 × 5 = 30  6 × 6 = 36
7 × 1 =  7  7 × 2 = 14  7 × 3 = 21  7 × 4 = 28  7 × 5 = 35  7 × 6 = 42  7 × 7 = 49
8 × 1 =  8  8 × 2 = 16  8 × 3 = 24  8 × 4 = 32  8 × 5 = 40  8 × 6 = 48  8 × 7 = 56  8 × 8 = 64
9 × 1 =  9  9 × 2 = 18  9 × 3 = 27  9 × 4 = 36  9 × 5 = 45  9 × 6 = 54  9 × 7 = 63  9 × 8 = 72  9 × 9 = 81
```

图3-10 九九乘法表

Figure 3-10 Multiplication Table of 9

算法分析：在九九乘法表中，每一行的乘法算式个数随着当前行数的行号在变化，所以一般来说，可构造外层循环控制行数，内层循环控制列数的嵌套循环。以前3行为例，第1行，打印1个乘法算式，第2行，打印2个乘法算式，第3行，打印3个乘法算式，以此类推，外层循环执行9次，因为有9行需要打印，而每一行打印的列数，由当前行数决定。参考代码如下：

Algorithm analysis: In the multiplication table of 9, the number of multiplication equations in each row changes with the current row number. Therefore, in general, a nested loop is constructed, where the outer loop controls the row number and the inner loop controls the column number. Taking the first three rows as an example, in the first row, one multiplication equation is printed. In the second row, two multiplication equations are printed. In the third row, three multiplication equations are printed, and so on. The outer loop executes 9 times because there are 9 rows to print, and the number of columns to print in each row is determined by the current row number. The reference code is as follows:

```python
#The outer loop controls the number of rows
for i in range(1, 10):
    #The inner loop controls the number of columns
    for j in range(1, i+1):
        #Print the product and use the end parameter to avoid line break
        #">" means right alignment, 2 represents a width of two characters, d represents output in integer format.
        print("{} × {} = {:>2d}".format(i,j,i*j), end=" ")
    #After printing each line, move to a new line
    print()
```

3-16

[Example 3-17] The Problem of Hundred Coins Hundred Chickens is a classic ancient Chinese mathematical problem. A farmer has 100 coins and needs to buy 100 chickens. Roosters cost 5 coins each, hens cost 3 coins each, and chicks cost 1 coin for three. How can the farmer make sure to spend all the money and buy exactly 100 chickens, with the requirement that the quantities of roosters, hens, and chicks are all nonzero.

Algorithm analysis: This problem can be solved using nested loops to iterate over the quantities of roosters and hens, the number of chicks is 100 minus the number of roosters and the number of hens. Here is a reference code that solves the Problem of Hundred Coins Hundred Chickens using nested loops.

```
for cock in range(1,21):            #The number of roosters is between 1 and 20
    for hen in range(1,34):         #The number of hens is between 1 and 33
        chick=100-cock-hen          #The number of chicks
        if cock*5+hen*3+chick/3==100:
            print(cock,hen,chick)   #Print the combinations that meet the criteria
```

3-17

3.4.8 循环中的 else 子句
3.4.8 The else Clause within a Loop

在 Python 中，for 循环和 while 循环都可以与 else 子句搭配使用，当循环正常结束时（没有通过 break 语句中断循环或遇到异常时），else 子句将会执行。如果循环被 break 语句中断，则 else 子句不会执行。

In Python, both for loops and while loops can be combined with an else clause, which will be executed when the loop completes normally(without being interrupted by a break statement or encountering any exceptions). If the loop is interrupted by a break statement, the else clause will not be executed.

```
for i in range(1, 4):
    print(f"Checking number: {i}")
    if i==3:
        print("Found it!")
        break
else:
    print("Loop ended without finding number 3")
```

在这个例子中，当 i 等于 3 时，break 语句会被执行，因此 else 子句不会执行。但是，如果我们将条件改为 if i == 5（一个不在

In this example, when i is equal to 3, the break statement will be executed, and therefore the else clause will not be executed. However, if we change the condition to if i ==5(a number that is not within the loop

循环范围内的数），则循环会正常结束，else 子句将会执行。

range), the loop will complete normally, and the else clause will be executed.

```
i=1
while i<4:
    print(f"Checking number: {i}")
    i+=1
    if i==4:
        print("Reached number 4")
        break
else:
    print("Loop ended without reaching number 4")
```

在这个例子中，当 i 增加到 4 时，break 语句会被执行，因此 else 子句不会执行。如果我们将条件改为 if i ==5，因为 i 的值最大为 4，所以 break 语句不会执行，循环正常结束，else 子句将会执行。

【例 3-18】判断正整数 n 是否是素数，如果是，则输出"n is a prime number"，否则输出"n is not a prime number"。

算法分析：在这个例子中，如果用户输入的正整数 n 能够被 2 到 sqrt(n) 之间的任何整数整除，则输出"n is not a prime number"，并且通过 break 语句跳出循环。如果循环正常结束（没有触发 break），那么 else 子句将会执行，输出"n is a prime number"。

In this example, when i increases to 4, the break statement will be executed, therefore the else clause will not be executed. If we change the condition to if i ==5, since the maximum value of i is 4, the break statement will not be executed, the loop will complete normally, and the else clause will be executed.

[Example 3-18] Check whether the positive integer n is a prime number. If it is, output "n is a prime number". Otherwise, output "n is not a prime number".

Algorithm analysis: In this example, if the positive integer n entered by the user can be divided by any integer between 2 and the square root of n, then "n is not a prime number" will be outputted and the loop is exited using the break statement. If the loop completes normally (without triggering a break), then the else clause will be executed, and "n is a prime number" will be outputted.

```
import math
n=int(input("Please enter a positive integer: "))
#Checking all integers from 2 to the square root of n
for i in range(2, int(math.sqrt(n))+1):
    if n % i==0:          #If n is divisible by i, n is not a prime number
        print(f"{n} is not a prime number ")
        break             #Break out of the loop
else:   #The for loop ended normally without triggering a break
    print(f"{n} is a prime number ")
```

3-18

与例 3-14 不同的是，这里使用 math.sqrt (n) 来计算 n 的平方根，并将其转换为整数，以优化缩减循环的次数，这是因为一个合数必然有一个不超过它平方根的因数。通过这种方式，可以减少循环次数，提高程序执行效率。

Here, what differs from Example 3-14 is math.sqrt(n) is used to calculate the square root of n and convert it to an integer in order to optimize and reduce the number of iterations. This is because a composite number must have a factor that does not exceed its square root. By doing this, it can reduce the number of iterations and improve the efficiency of program execution.

3.5 异常处理
3.5 Exception Handling

在 Python 中，异常处理是通过 try-except-else-finally 结构来实现的。这种结构允许处理可能在程序运行时发生的异常，从而避免程序崩溃。

In Python, exception handling is implemented using the try-except-else-finally structure. This structure allows for handling exceptions that may occur during program execution, preventing the program from crashing.

try-except-else-finally 的基本语法如下：

The basic syntax of try-except-else-finally is as follows:

```
try:
    #Code block 1 to be attempted for execution
    Code block 1
except ExceptionType1:
    #Code block 2 to be executed when an exception of type "ExceptionType1" occurs
    Code block 2
except ExceptionType2:
    #Code block 3 to be executed when an exception of type "ExceptionType2" occurs
    Code block 3
else:
    #If Code block 1 in the try block does not raise any exceptions, then execute Code block 4 in the else block
    Code block 4
finally:
    #Code block 5 in the finally block will be executed regardless of whether an exception occurs or not
    Code block 5
```

（1）try 模块：首先，try 模块中的代码被执行。如果这段代码执行顺利，没有发生任何异常，那么程序将继续执行 else 模块（如果存在的话），然后执行 finally 模块（如果存在的话）。

(1) try block: First, the code in the try block is executed. If this code runs smoothly without any exceptions, the program will continue executing the code in the else block(if it exists), and then proceed to execute the code in the finally block(if it exists).

107

（2）except 模块：如果在 try 模块的代码执行过程中发生了异常，那么程序将立即跳到相应的 except 模块中。Python 会查找与引发的异常类型相匹配的 except 模块。如果找到了匹配的模块，则该模块中的代码将被执行。

（3）else 模块：else 模块是可选的，它包含了当 try 模块中的代码没有引发任何异常时应该执行的代码。如果 try 模块中有异常发生，那么 else 模块将被跳过。

（4）finally 模块：finally 模块也是可选的，它包含了无论是否发生异常都应该执行的代码。这对于资源清理操作，如关闭文件、释放锁等非常有用，因为这些操作通常需要在程序结束前无条件地执行。

【例 3-19】两个数做除法运算，注意，数字 0 不能作为除数，程序要能够对异常输入进行处理，使程序不会因为除数是 0 而崩溃。

(2) except block: If an exception occurs during the execution of the code in the try block, the program will immediately jump to the corresponding except block. Python will search for an except block that matches the raised exception type. If a matching block is found, the code within that block will be executed.

(3) else block: The else block is optional and contains code that should be executed when no exceptions are raised in the try block. If an exception occurs in the try block, the else block will be skipped.

(4) finally block: The finally block is also optional and contains code that should be executed regardless of whether an exception occurs or not. This is particularly useful for performing cleanup operations, such as closing files or releasing locks, as these operations typically need to be executed unconditionally before the program ends.

[Example 3-19] Perform division operation on two numbers, ensuring that the number 0 cannot be used as the divisor. The program should be able to handle exceptional inputs and prevent the program from crashing due to division by zero.

```
dividend = int(input("Please enter the dividend: "))
divisor = int(input("Please enter the divisor: "))
#Enter 0 to trigger the ZeroDivisionError exception.
try:
    result = dividend / divisor
    print(f"The result is: {result}")
except ZeroDivisionError:
    #The except block will be executed when the divisor is 0
    print("Error: The divisor cannot be zero!")
```

3-19

在这个例子中，定义两个变量 dividend（被除数）和 divisor（除数），用户可以故意将 divisor 设置为 0。然后，使用 try 语句来尝试执行除法操作。由于除数是 0，这会导致 ZeroDivisionError 异常被抛出。

In this example, two variables, dividend(the number to be divided) and divisor(the number to divide by), users can intentionally set the divisor to 0. Then, a try statement is used to perform the division operation. Due to the divisor being 0, this will result in the ZeroDivisionError exception being raised.

第3章 控制结构
Chapter 3　Control Structure

except 语句用于捕捉这个特定的异常。当 ZeroDivisionError 异常发生时，程序控制流会立即跳转到 except 模块，并执行其中的代码。在这个例子中，只是简单地打印出一条错误消息，告诉用户除数不能为 0。如果除数不为 0，那么程序会正常执行并输出商。

表 3-1 列出了 Python 中常见异常名称及其描述，Python 中还有许多其他类型的异常。从 Python 2 到 Python 3，一些异常（如 IOError）已经被更具体的异常类型所替代。在编写代码时，建议查阅 Python 的官方文档以获取最准确和最新的信息。

The except statement is used to catch this specific exception. When the ZeroDivisionError exception occurs, the program's control flow will immediately jump to the except block and execute the code within it. In this example, it simply prints an error message informing the user that the divisor cannot be zero. If the divisor is not equal to 0, the program will execute normally and output the quotient.

The table 3-1 below lists common exception names and their descriptions in Python. There are many other types of exceptions in Python. Some exceptions, such as IOError, have been replaced by more specific exception types from Python 2 to Python 3. When writing code, it is recommended to consult the official Python documentation for the most accurate and up-to-date information.

表 3-1　常见异常名称及其描述
Table 3-1　Common Exception Names and Descriptions

异常名称	描述	description
Exception	所有异常的基类，当使用 except 而不指定特定异常类型时，会捕获所有 Exception 的子类异常	The base class for all exceptions, when using 'except' without specifying a specific exception type, it will catch all sub-classes of Exception
AttributeError	当试图访问的对象属性或方法不存在时引发	The exception raised when trying to access a nonexistent object attribute or method
IndexError	当使用序列中不存在的索引时引发，例如访问列表的索引超出范围	The exception raised when attempting to access an index that doesn't exist in a sequence, such as accessing an index beyond the range of a list
KeyError	当使用字典中不存在的键时引发	The exception raised when trying to access a key that doesn't exist in a dictionary
NameError	当尝试访问一个还未被赋予对象的变量时引发	The exception raised when attempting to access a variable that has not been assigned an object yet
SyntaxError	当 Python 解释器在解析代码时遇到语法错误时引发	The exception raised when the Python interpreter encounters a syntax error while parsing the code

109

续表

异常名称	描述	description
TypeError	当对类型的操作或函数应用于不适当类型的对象时引发	The exception raised when operations or functions are applied to inappropriate types of objects
ValueError	当内置操作或函数应用于正确类型的对象，但该对象使用不合适的值时引发	The exception raised when built-in operations or functions are applied to the correct type of object, but that object has an inappropriate value
ZeroDivision-Error	当除法或取模运算的第二个参数为0时引发	The exception raised when the second argument of a division or modulus operation is 0
OverflowError	运算结果超出了所允许的范围，这通常与整数的大小限制有关	The exception raised when the result of an operation exceeds the allowed range, usually associated with the size limits of integers
TabError	混用Tab和空格进行缩进	The exception raised when mixing tab and space for indentation
ImportError	模块导入失败	The exception raised when a module fails to import
ModuleNot-FoundError	模块不存在	The exception raised when a module does not exist

3.6 本章小结
3.6 Summary of This Chapter

本章讲解了程序的3种基本结构，即顺序结构、分支结构、循环结构，以及异常处理，主要内容如下。

（1）分支结构中的多分支结构最为常见，在多分支结构中，if 和 elif 不可以省略，且 if 和 elif 后必须有条件表达式，当表达式为真时，执行相应的缩进语句块。

else 后面无条件表达式，直接以冒号结束，只有当 else 前面所有

This chapter explains the three basic structures of programming, namely sequential structure, conditional structure, loop structure, as well as exception handling. The main contents are as follows.

(1)In conditional structures, multi-branching structure is the most commonly used. In a multi-branch structure, if and elif cannot be omitted, and there must be a conditional expression after if and elif. When the expression is true, the corresponding indented block of statements is executed.

else does not have a conditional expression after it, it is directly followed by a colon. Only when all the con-

条件表达式为假时，才执行 else 后面的缩进语句块。

（2）break 的作用是满足某个条件时，触发 break 语句，此时提前结束循环，接着执行循环外的语句。

而 continue 的作用是当满足某个条件时，continue 执行，跳过当前循环下 continue 后面剩余的语句，提前进入下一次循环。

（3）在 Python 中，for 循环用于遍历序列（如 range()函数、字符串、列表、元组等），执行固定次数的循环操作，又称为遍历循环。而 while 循环在满足条件时反复执行代码块。二者主要区别在于：for 循环适用于循环次数确定的场景；while 循环适用于循环次数不确定，达到某个临界条件时结束循环的场景。while 还可以构造出无限循环，到达一定循环次数时，使用 break 终止循环。else 子句与 for 和 while 循环结合时，仅在循环正常完成之后执行。

（4）range 是一种数据类型，表示不可变的等差数列，range()函数产生的可迭代对象通常用于控制 for 循环的执行次数，range(start, stop, step)，可获得从 start 到（stop-1）、步长为 step 的整数数列。只有 start 参数和 step 参数可以省略，start 省略默认从 0 开始，step 省略默认步长为 1。

（5）pass 语句是一个空语句，不执行任何操作，一般作为占位使用，为了让程序结构完整。

（6）Python 程序无法正常运行

ditional expressions before else are false, the indented block of statements after else is executed.

(2)The purpose of break is to trigger the break statement when a certain condition is met, causing the loop to end prematurely and executing the statements outside the loop.

On the other hand, the purpose of continue is to execute when a certain condition is met, skip the remaining statements after continue in the current loop, and proceed to the next iteration of the loop.

(3)In Python, the for loop is used to iterate over a sequence(such as the range() function, strings, lists, tuples, etc.) and performs a fixed number of loop operations. It is also known as an iterative loop. On the other hand, the while loop repeatedly executes a code block as long as a condition is satisfied. The main difference between the two is that the for loop is suitable for situations where the number of iterations is known in advance, the while loop is suitable for situations where the number of iterations is uncertain and the loop terminates when a certain condition is met. The while loop can also create an infinite loop, in which case one can use the break statement to stop the loop after a certain number of iterations. When use else clause with for loops and while loops, the else clause is executed only when the loop completes normally.

(4)The range is a data type that represents an immutable arithmetic sequence. The range() function generates an iterable object that is commonly used to control the execution of for loops. It generates a sequence of integers from start to(stop-1), with a step size of step. Only the start and step parameters can be omitted. If start is omitted, it defaults to 0, and if step is omitted, it defaults to 1.

(5) The pass statement is an empty statement that does nothing. It is typically used as a placeholder to ensure the completeness of the program's structure.

(6)When a Python program encounters an error and

时，可能会发生异常，使用 try-except-else-finally 结构捕捉异常、处理异常，使程序可以正常运行，加强程序的健壮性。

cannot run normally, exceptions may occur. The try-except-else-finally structure is used to catch and handle exceptions, allowing the program to run without interruption and enhancing its robustness.

3.7 本章练习
3.7 Exercise for This Chapter

3.1 根据一个学生的分数来评定其等级，规则如下：90 分及以上为 A 级，80 分及以上但不足 90 分为 B 级，70 分及以上但不足 80 分为 C 级，60 分及以上但不足 70 分为 D 级，60 分以下为 E 级，输入成绩则输出对应等级。

3.1 To evaluate a student's grade based on their score, the rules are as follows: A grade for scores of 90 and above, B grade for scores of 80 and above but less than 90, C grade for scores of 70 and above but less than 80, D grade for scores of 60 and above but less than 70, and E grade for scores below 60. Input the score and output the corresponding grade.

3.2 实现分段函数计算，如表 3-2 所示，输入 x 的值，输出对应分段函数内 y 的值。

3.2 Implement piecewise function calculation as shown in the Table 3-2. Input the value of x and output the corresponding value of y within the piecewise function.

表 3-2 分段函数
Table 3-2 Piecewise function

x	y
x<0	0
0<=x<5	x
5<=x<10	3x-5
10<=x<20	0.5x-2
20<=x	0

3.3 从键盘输入任意一个正整数，编程计算该数的阶乘，例如：输入 5，阶乘结果为 120。

3.3 Input any positive integer from the keyboard, then calculate the factorial of that number. For example, if the input is 5, the factorial result is 120.

3.4 打印出 1~100 的所有奇数，使用 while 循环结构来实现。

3.4 Print all odd numbers between 1 and 100 using a while loop structure.

3.5 计算如下多项式的和：20+21+22+23+24+…+263。

3.5 Calculate the sum of the following polynomial: 20+21+22+23+24+…+263.

3.6 输出 200~300 的所有素数，每 5 个数为一行输出。

3.6 Output all prime numbers between 200 and 300, output every five numbers per line.

3.7 输出 500 以内所有完数

3.7 Output all perfect numbers within 500.(For ex-

(例如，6 是一个完数，6 的因子 1、2、3（不包括 6）之和为 6，与其本身数字 6 相等，所以 6 是一个完数）。

3.8 打印图 3-11 所示图形，使用循环嵌套结构来实现。

ample, 6 is a perfect number because its factors 1, 2, 3 (excluding 6) add up to 6, which equals the number itself, making it a perfect number).

3.8 Print the graphics shown in Figure 3-11 using nasted loop structure.

******
######
******
######

图 3-11　图形打印
Figure 3-11　Graphic Printing

本章练习题参考答案

第 4 章 函数
Chapter 4　Functions

　　程序员在面对复杂的编程挑战时，常常采取一种策略，即将整个复杂的程序拆分成一系列更小、更易于管理的子问题。对于每一个子问题，程序员会编写一个独立的模块，这样可以大大简化编程的复杂性。最终，这些模块就像积木一样被组合在一起，形成完整的程序。这种将大问题分解为小问题，并分别解决，最后整合的程序设计方法，我们称之为模块化的程序设计方法，它有效地提高了程序的可读性、可维护性和可重用性。

　　在 Python 中，子问题是通过函数的形式实现的。函数就是组织好的，可以重复使用的，用来实现某个功能的代码段，如图 4-1 所示。使用函数可以降低代码重复率，提升代码的整洁度，使代码更容易理解和重用。

　　When facing complex programming challenges, programmers often adopt a strategy, that is, to break down the entire complex program into a series of smaller, more manageable subproblems. For each subproblem, the programmer will write an independent module, which can greatly simplify the complexity of programming. Ultimately, these modules are combined together like building blocks to form a complete program. This approach of breaking down a big problem into smaller ones, solving them separately, and then integrating them is known as modular programming, which effectively enhances the readability, maintainability, and reusability of the program.

　　In Python, subproblems are implemented in the form of functions. A function is an organized, reusable piece of code designed to perform a specific task, as shown in Figure 4-1. By using functions, code repetition can be reduced, code cleanliness can be improved, making the code easier to understand and reuse.

第4章 函数
Chapter 4 Functions

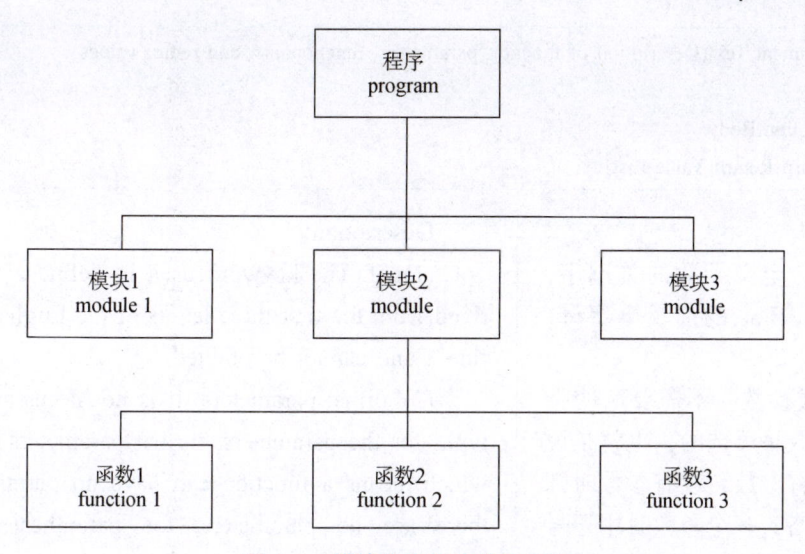

程序
program

模块1
module 1

模块2
module

模块3
module

函数1
function 1

函数2
function 2

函数3
function 3

图 4-1 模块化程序设计示意图
Figure 4-1 Modular Program Design Diagram

4.1 普通函数
4.1 Regular Functions

在前面的章节中我们已经深入学习了 Python 自带的内置函数，例如 print()函数用于输出信息，input()函数用于从键盘上获取用户输入的数据，pow()函数用于计算一个数的幂等。然而，这些内置函数可能无法涵盖所有用户的实际需求。在这种情况下，为了满足特定的功能需求，就需要自己编写函数来实现特定功能。

In the previous chapters, we have delved into studying the built-in functions in Python, such as the print() function used for outputting information, the input() function for obtaining user input from the keyboard, and the pow() function for calculating the power of a number. However, these built-in functions may not cover all the actual needs of users. In such cases, to meet specific functional requirements, we need to write our own functions to implement the desired functionality.

4.1.1 函数的定义
4.1.1 Function Definition

函数是一段被命名的代码块，用于执行特定的任务，使程序更模块化、更易于理解和维护。在 Python 中，函数必须先定义再使用。定义函数的语法格式如下：

A function is a named block of code used to perform a specific task, making the program more modular, easier to understand and maintain. In Python, functions must be defined before they are used. The syntax for defining a function is as follows:

```
def  Function Name([Formal Parameter List]):
    '''
```

115

```
Comment Text(Description of function parameters, functionality, and return values)
'''
Function Body
[return Return Value List]
```

说明：

（1）def：定义函数的关键字，取英文单词 define 的前 3 个字母，不可省略。

（2）形式参数：不需要声明类型，形式参数是可选的，也就是说函数可以没有参数，没有参数时圆括号不能省略，多个参数时用逗号分隔。

（3）形式参数圆括号后的":"不能省略。

（4）文档注释：一个可选的字符串，用于解释函数的作用、参数和返回值。

（5）函数体：是包含实际执行的代码的块，需要和 def 关键字保持一定的空格缩进。函数体可以使用 return 语句返回值，即返回函数的执行结果。

【例 4-1】定义函数示例。定义一个名为 add 的函数，形式参数 a、b 用于接收两个数值型数据，函数功能为计算 a 和 b 的和，通过 return 语句返回计算结果。

Description:

(1)def: The keyword used to define a function, derived from the first three letters of the English word "define", and cannot be omitted.

(2)Formal parameters: It is not necessary to declare types for the parameters, formal parameters are optional, which means a function can have no parameters. When there are no parameters, the parentheses cannot be omitted, and when there are multiple parameters, they are separated by commas.

(3)The colon ":" after the formal parameter parentheses cannot be omitted.

(4) Documentation annotation: An optional string used to explain the purpose of the function, its parameters, and return values.

(5)Function body: It is a block of code that contains the actual execution code, and it should be indented with a certain number of spaces after the def keyword. The function body can use the return statement to return a value, which is the result of the function execution.

[Example 4-1] Function definition example. Define a function named add with formal parameters a and b to accept two numerical values. The function's purpose is to calculate the sum of a and b, and return the result using the return statement.

```
def add(a,b):
    '''Calculate the sum of two numbers'''
    c=a+b
    return c
```

4-1

4.1.2 函数的调用
4.1.2 Function Call

函数调用是指执行一个已经定义的函数，并传递给它必要的参数

Function call refers to executing a function that has been defined and passing the necessary arguments to it(if

Chapter 4　Functions

（如果有的话）。当函数被调用时，Python 会执行函数体中定义的代码块，并根据需要返回结果给函数调用方。函数调用的语法格式如下：

any). When a function is called, Python will execute the code block defined in the function body and return the result to the function caller as needed. The syntax for a function call is as follows:

```
Function Name([Actual parameter list])
```

说明：

（1）调用函数的函数名要与定义一致。

（2）函数调用中的参数为实际参数，即从主程序向定义的函数传递的参数值。需要注意的是，函数定义中的形参和实参要保持个数、顺序一致。

【例 4-2】调用例 4-1 定义的函数示例。

Description:

(1)The function name used in the function call must match the defined function name.

(2)The parameters in a function call are actual parameters, meaning the values passed from the main program to the defined function. It is important to ensure that the number and order of parameters in the function definition match those in the function call.

[Example 4-2] Call the function defined in Example 4-1.

```
def add(a,b):
    '''Calculate the sum of two numbers'''
    c=a+b
    return c
x,y=eval(input())
z=add(x,y)    #Call the add() function
print(f'{x}+{y}={z}')
```

4-2

（3）函数必须先定义后调用，若在定义之前调用，则会报错。

【例 4-3】先定义后调用。

(3)Functions must be defined before they are called, otherwise, an error will occur if they are called before being defined.

[Example 4-3] Define first, then call.

```
def fun():
    print('Hello Python!')
fun()
```

4-3

输出：

Output:

```
Hello Python!
```

【例 4-4】先调用后定义。

[Example 4-4] Call first, define later.

```
fun()
```

117

```
def fun():
    print('Hello Python!')
```

输出: | Output:

```
Traceback(most recent call last):
    File "C:\test. py", line 1, in<module>
        fun()
NameError: name 'fun' is not defined
```

4-4

(4)若函数名被重复定义，则调用离它最近的函数定义。可以理解为后定义的函数覆盖了先定义的函数。

【例 4-5】函数重复定义。

(4)If a function name is redefined, the function call will refer to the closest function definition. This can be understood as the function defined later overriding the function defined earlier.

[Example 4-5] Function redefinition.

```
def f():         #Define function f()
    print('Hello World!')

def f():         #Define a function named f() again
    print('Hello Python!')

#Call the closest function definition when the function name is redefined
f()              #Call the function f() defined most recently
```

4-5

输出: | Output:

```
Hello Python!
```

4.1.3 函数的返回值
4.1.3 Return Value of a Function

在 Python 中，函数可以返回一个或多个值。返回的值可以是任何数据类型，包括整数、浮点数、字符串、列表、字典、元组等，甚至可以是另一个函数或类的实例。

返回值通过 return 语句将值带回给主调函数，其位置在函数体内。语法格式如下：

In Python, functions can return one or multiple values. The returned values can be of any data type, including integers, floats, strings, lists, dictionaries, tuples, etc, and even another function or an instance of a class.

The return statement is used to send back the values to the calling function, and it appears within the function body. The syntax is as follows:

```
return expression1, expression2, …, expressionN
```

说明：

(1)函数可以有返回值，即调用函数后获取的值，也可以没有返回值。

(2)当函数中无 return 语句或者 return 语句后面无表达式时，函数返回值为 None。

(3)一条 return 语句可以同时返回多个值，用逗号隔开。

【例 4-6】函数有多个返回值示例(实参与形参个数一致)。

Description：

(1)Functions can have return values, which are the values obtained when calling a function, or they can have no return value.

(2)If a function does not have a return statement or the return statement has no expression following it, the function's return value is None.

(3) A single return statement can return multiple values separated by commas.

[Example 4-6] Function with multiple return values (Number of actual parameters matches number of formal parameters).

```
def  func(a,b):
    return a+b,a-b,a*b,a/b        #Four return values
a,b,c,d=func(10,20)
print(a,b,c,d)                    #Four variables receive return values
```

4-6

输出：

Output:

```
30 -10 200 0.5
```

(4)主程序也可以只有一个变量接收多个返回值，此变量为元组类型。

【例 4-7】函数有多个返回值时，用一个变量接收它们。

(4) The main program can also have only one variable to receive multiple return values, this variable is in the form of a tuple type.

[Example 4-7] Function has multiple return values and one variable receiving them.

```
def  func(a,b):
    return a+b,a-b,a*b,a/b        #Four return values
t=func(10,20)                     #A variable receives 4 return values
print(t)
```

4-7

输出：

Output:

```
(30, -10, 200, 0.5)
```

4.1.4 扩展例题
4.1.4 More Examples

【例 4-8】编写函数 max()，形参为 a 和 b，用户输入两个数，通过调用该函数，求两个数的最

[Example 4-8] Write a function max() that takes parameters a and b. The user inputs two numbers, and by calling this function, the maximum of the two numbers is

119

大值。 | calculated.

```
def  max(a,b):                    #Define the function
    if a>=b:
        return a
    else:
        return b
    #return a if a>=b else b       #conditional expression

m=int(input("Enter an integer m: "))
n=int(input("Enter an integer n: "))
print(max(m,n))                    #Call the function
```

4-8

输出： | Output:

```
Enter an integer m: 20
Enter an integer n: 12
20
```

【例 4-9】编写函数 calSum()，形参为 n，用户输入正整数，调用该函数，求 1 到 n 的累加和（包括 n）。 | [Example 4-9] Write a function calSum() that takes parameter n. The user inputs a positive integer, and by calling this function, the sum of numbers from 1 to n (including n) is calculated.

```
def  calSum(n):
    s=0
    for i in range(n+1):
        s+=i
    return s

m=int(input("Enter an integer m: "))
print('sum=',calSum(m))    #Calling the function calSum()
```

4-9

输出： | Output:

```
100
sum= 5050
```

4.2 函数的参数
4.2 Function Parameters

在 Python 中，函数参数分为形式参数（形参）和实际参数（实 | In Python, function parameters are divided into formal parameters and actual parameters.

参）两种。

（1）形参是函数定义中括号内的参数名。它们定义了函数期望接收的值的类型和数量。在函数体内部，可以使用这些形参名来引用传递进来的值。

（2）实参是调用函数时传递给函数的实际值。它们可以是常量、变量、表达式或其他函数调用的结果。

【例 4-10】形参和实参示例。

(1)Formal parameters are the parameter names specified within the function definition's parentheses. They define the type and number of values the function expects to receive. Within the function body, one can use these parameter names to reference the values passed in.

(2)Actual parameters are the actual values passed to the function when it is called. They can be constants, variables, expressions, or the results of other function calls.

[Example 4-10] Example of formal parameters and actual parameters.

```
def greet(name, age):     #name and age are formal parameters.
    print(f"Hello! My name is {name} and I'm {age} years old.")

greet('Bob',20)           #'Bob', 20 are actual parameters
```

4-10

输出：

Output：

Hello! My name is Bob and I'm 20 years old.

4.3 函数的参数传递
4.3 Parameter Passing in Functions

在 Python 中，参数是一种用于在调用函数时向函数内部传递数据的方式。Python 支持多种类型的参数，使得函数更加灵活和可重用。当函数的定义中存在多个参数时，其参数传递的形式有位置传递、关键字传递、默认值传递、可变长参数传递和解包裹传递等。

In Python, parameter is a way to pass data into the function during a function call. Python supports various types of parameters, making functions more flexible and reusable. When a function has multiple parameters in its definition, the parameter passing can take the form of positional passing, keyword passing, default value passing, variable-length parameter passing, and unpacking passing.

4.3.1 位置传递
4.3.1 Positional Passing

位置传递是最常用的传参方式，参数的位置固定，参数传递时按照形参定义的顺序提供数据。在调用函数时，必须按照与形参相同的顺序提供值。

Positional passing is the most commonly used way of passing arguments, where the positions of the parameters are fixed. When passing arguments, data is provided in the order defined by the formal parameters. When calling a function, values must be provided in the same

121

【例 4-11】位置传递示例。

[Example 4-11] Example of positional passing.

```
def func(name,age,city):
    '''Personal information, parameters are name, age, and city'''
    print("My name is {},{} years old, from {}.".format(name,age,city))
Name="LiMing"
Age=18
City="Anshan"
func(Name,Age,City)    #The actual parameters are in the same position as the formal parameters
```

输出：

Output:

My name is LiMing,18 years old, from Anshan.

4.3.2 关键字传递
4.3.2 Keyword Passing

关键字传递是指调用函数时，通过参数名来指定参数的值，而不依赖于参数的顺序。

Keyword passing refers to specifying the values of parameters by parameter names when calling a function, rather than relying on the order of the parameters.

【例 4-12】关键字传递示例。

[Example 4-12] Example of keyword passing.

```
def func(name,age,city):
    ''' Personal information, parameters are name, age,city'''
    print("My name is {},{} years old, from {}.".format(name,age,city))

Name="LiMing"
Age=18
City="Anshan"
#Keyword passing does not depend on the position of the parameters
func(age=Age,city=City,name=Name)
```

输出：

Output:

My name is LiMing,18 years old, from Anshan.

关键字传递与位置传递可以同时使用，但按位置传递的参数要放在按关键字传递的参数前面，否则，编译器无法知道关键字外的参数出现的顺序。

Keyword passing and positional passing can be used together, but positional arguments must be placed before keyword arguments. Otherwise, the compiler will not be able to determine the order of parameters appearing after the keywords.

【例 4-13】关键字传递和位置传递同时存在的示例。

[Example 4-13] An example where keyword passing and positional passing are used simultaneously.

```
def func(name,age,city):
    '''Personal information, parameters are name, age, city'''
    print("My name is {},{} years old, from {}.".format(name,age,city))

Name="LiMing"
Age=18
City="Anshan"
#name is passed by positional passing, the other parameters are passed by keyword passing
func(Name,city=City,age=Age)
```

输出： | Output:

My name is LiMing,18 years old, from Anshan.

4.3.3 默认值传递
4.3.3 Default Value Passing

默认值传递是指定义函数时，直接给形参赋予默认值。如果调用函数时没有提供该参数的值，则使用默认值。需要注意的是，当函数有多个参数时，默认值参数必须放在后面，非默认值参数放在前面。

Default value passing refers to assigning default values directly to formal parameters when defining a function. If a value of a parameter is not provided when calling a function, the default value is used. It is important to note that when a function has multiple parameters, default value parameters must be placed at the end, with non-default value parameters at the beginning.

【例 4-14】默认值传递的函数定义示例。

[Example 4-14] The example of a function definition where the default value passing is used.

```
#The default value parameters are placed at the end
def func1(name,age,city='Anshan'):
    print("My name is {},{} years old, from {}.".format(name,age,city))

#Multiple default value parameters can be placed afterwards
def func2(name,age=18,city='Anshan'):
    print("My name is {},{} years old, from {}.".format(name,age,city))

#Positional parameters should be placed before default value parameters
def func3(name='Liming',age,city='Anshan'):
    print("My name is {},{} years old, from {}.".format(name,age,city))
#"SyntaxError: non-default argument follows default argument"
```

示例中的前两个函数定义正确，第三个函数定义出现语法错误，因为位置参数 age 放在了默认

The first two function definitions in the example are correct, but the third function definition contains a syntax error because the positional parameter "age" is placed af-

值参数 name 的后面，运行时会报"SyntaxError: non-default argument follows default argument"错误。

【例 4-15】默认值参数传递示例。

ter the default value parameter "name". This will result in a "SyntaxError: non-default argument follows default argument" error at runtime.

[Example 4-15] Default value passing example.

```
#The default value parameter is placed at the end
def func(name,age,city='Anshan'):
    print("My name is {},{} years old, from {}.".format(name,age,city))
Name="LiMing"
Age=18
City="Beijing"
func(Name,age=Age)
func(Name,Age,City)
```

4-15

输出：

Output:

```
My name is LiMing,18 years old, from Anshan.
My name is LiMing,18 years old, from Beijing.
```

在本例中，当默认值参数 city 无实参传入时，使用默认值 Anshan；有实参传入时，实参的值 Beijing 被使用。

In this example, when the default parameter city is called without an actual argument, the default value Anshan is used. When an actual argument is passed, the value of the argument Beijing is used.

4.3.4 可变长参数传递
4.3.4 Variable-length Parameter Passing

前面介绍的几种参数传递，需要在定义函数时就确定好需要多少个参数。但是有的时候我们无法知道参数个数，这时可以使用可变长参数传递。在 Python 中，有两种方式实现可变长参数传递，分别是使用 *args 和 **kwargs 来实现，它们用于接收可变数量的参数。

（1）*args 用于传递非关键字参数，它将接收到的所有位置参数组成一个元组。

（2）**kwargs 用于传递关键字参数，它将接收到的所有关键字参数组成一个字典。

In the previous introduced forms of parameter passing, the number of parameters needed must be determined when defining the function. However, there are situations where the number of parameters is unknown. In such cases, variable-length parameter passing can be used. In Python, there are two ways to implement variable-length parameter passing, which are using *args and **kwargs to receive a variable number of arguments.

(1)*args is used to pass non-keyword arguments, it collects all the positional arguments into a tuple.

(2)**kwargs is used to pass keyword arguments, it collects all the keyword arguments into a dictionary.

【例 4-16】可变长参数传递示例。

[Example 4-16] Variable-length parameter passing example.

```
def sum_numbers(*num):
    return sum(num)

def fun(**kwargs):
    for key, value in kwargs.items():
        print(f"{key} = {value}")

print(sum_numbers(1, 2, 3, 4))
fun(a=3, b=4, c=5)
```

输出：

Output:

```
10
a = 3
b = 4
c = 5
```

需要注意的是，*args 和 **kwargs 可以同时在函数定义中使用，但它们必须位于函数参数列表的最后。

这是因为 Python 解释器在处理函数参数时，会先处理位置参数和关键字参数，然后将剩余的位置参数和关键字参数分别传递给 *args 和 **kwargs。

It is important to note that *args and **kwargs can be used simultaneously in a function definition, but they must be placed at the end of the function parameter list.

This is because the Python interpreter processes positional and keyword arguments first when handling function parameters, and then the remaining positional and keyword arguments are passed to *args and **kwargs respectively.

4.3.5 解包裹传递
4.3.5 Unpacking Passing

在 Python 中，解包裹是一种将可迭代对象(如列表、元组或字典)的元素或键值对赋值给变量的技术。这在函数参数传递时特别有用，尤其是当想要将一个序列的元素作为单独的参数传递给函数时。

对于非关键字参数，可以使用 * 来解包裹一个序列，将其元素作为单独的参数传递给函数。对于关

In Python, unpacking is a technique that involves assigning the elements or key-value pairs of an iterable object(such as a list, tuple, or dictionary) to variables. This is particularly useful in function parameter passing, especially when one wants to pass the elements of a sequence as individual parameters to a function.

For non-keyword arguments, one can use * to unpack a sequence and pass its elements as separate arguments to a function. For keyword arguments, one can use ** to un-

键字参数，可以使用 ** 来解包裹一个字典，将其键值对作为关键字参数传递给函数。

【例4-17】解包裹参数传递示例。

pack a dictionary and pass its key-value pairs as keyword arguments to a function.

[Example 4-17] Unpacking parameter passing example.

```
def fun(a, b, c):
    print(a, b, c)
args = [1, 2, 3]
fun(*args)
kwargs = {'a': 4, 'b': 5, 'c': 6}
fun(**kwargs)
```

4-17

输出：

Output:

1 2 3
4 5 6

4.4 变量的作用域
4.4 Scope of Variables

变量的作用域就是指变量的有效范围，即能够在哪些范围内访问到它。有的变量可以在整段代码的任意位置使用，有的变量只能在函数内部使用。

变量的作用域由定义变量的位置决定。在 Python 中，根据作用域的不同，可将变量分为局部变量和全局变量。

The scope of a variable refers to the extent to which the variable is accessible, or where it can be accessed. Some variables can be used anywhere in the code, while others can only be used within a specific function.

The scope of a variable is determined by where the variable is defined. In Python, based on the scope, variables can be categorized into local variables and global variables.

4.4.1 局部变量
4.4.1 Local Variables

局部变量又称内部变量，是在函数或方法内部定义的变量，作用域仅限于函数内部，在函数外部不可以使用。形参被视为局部变量。当调用函数时，Python 会为局部变量分配临时的存储空间，函数执行完后，这块临时存储空间即被释

Local variables, also known as inner variables, are variables defined within a function or method, with their scope limited only to that particular function, they cannot be accessed outside of the function. Formal parameters are considered local variables. When a function is called, Python allocates temporary storage space for local variables. Once the function completes its execution, this tem-

放，因而里面存储的变量也就无法使用了。

【例4-18】局部变量示例。

```
def fun():
    s='Hello Python!'
    print(s)
fun()
print(s)
```

输出：

```
Hello Python!
Traceback(most recent call last):
    File "C:\test.py", line 6, in<module>
        print(s)
NameError: name 's' is not defined
```

示例中，函数 fun() 中定义了局部变量 s，该变量只能在函数内部使用，在函数外部使用则会触发 NameError 异常。

所以局部变量仅可以在函数内使用或修改，不能在函数外使用或修改。

4.4.2 全局变量

全局变量又称外部变量，是在函数或方法外部定义的变量，也可以是在函数内部使用 global 关键字声明的变量。全局变量默认的作用域是整个程序，既可以在函数内部使用或修改，也可以在函数外部使用或修改。

【例4-19】全局变量示例。

```
global_var=3.14
def fun1():
    global_var=2.71
    print('fun1:', global_var)
```

porary storage space is released, making the variables stored within it inaccessible.

[Example 4-18] Example of local variables.

4-18

Output:

In the example provided, the function fun() defines a local variable "s", which can only be used within the function. Attempt to use this variable outside the function would result in a NameError exception.

Therefore, local variables can only be accessed or modified within the function where they are defined and cannot be used or modified outside the function.

4.4.2 Global Variables

Global variables, also known as external variables, are variables defined outside of functions or methods, or declared within a function using the "global" keyword. The default scope of global variables is the entire program, meaning they can be accessed or modified both inside and outside functions.

[Example 4-19] Example of global variable.

4-19

```
def fun2():
    global global_var
    global_var=2.71
    print('fun2:',global_var)
print('outer:', global_var)
fun1()
print('outer:', global_var)
fun2()
print('outer:', global_var)
```

输出： | Output:

```
outer: 3.14
fun1: 2.71
outer: 3.14
fun2: 2.71
outer: 2.71
```

示例中，在函数外部定义的变量 global_var 为全局变量，fun1() 函数中定义的变量 global_var 为局部变量，和全局变量重名，屏蔽了全局变量。在 fun1() 函数中，对 global_var 的访问和修改均是对局部变量 global_var 的操作，对全局变量 global_var 没有影响。在 fun2() 函数中，使用 global 关键字声明 global_var 为全局变量，对 global_var 的访问和修改均是对全局变量 global_var 的操作。

In the example provided, the variable global_var defined outside functions is a global variable. In the fun1() function, the variable global_var is defined as a local variable, sharing the same name with the global variable. This effectively shields the global variable, so any access or modification of global_var within fun1() operates on the local variable, not affecting the global variable. In fun2(), the global keyword is used to declare global_var as a global variable, allowing access and modification of global_var to directly impact the global variable.

4.5 匿名函数
4.5 Anonymous Functions

匿名函数是一个没有名字的临时使用的小函数，通过 lambda 关键字创建，也称为 lambda 函数。如果一个函数在程序中只被调用一次，那么可以使用匿名函数。匿名函数可以接收任意数量的参数，但只能有一个表达式，该表达式会被求值

An anonymous function, also known as a lambda function, is a temporary small function without a name created using the lambda keyword. If a function is called only once in a program, an anonymous function can be used. Anonymous functions can take any number of parameters but can only consist of a single expression, which is evaluated and returned. Therefore, they are typi-

并返回。因此，它们通常用于简单的函数操作。其语法格式如下：

cally used for simple function operations. The syntax for an anonymous function is as follows:

lambda parameter list: expression

lambda 是匿名函数的关键字，冒号前面是参数列表，也可以没有参数。冒号后面是匿名函数的表达式，表达式只能占用一行。lambda 函数的定义其实就等价于：

The lambda is the keyword for anonymous functions, with the parameter list before the colon, the parameters can also be missing. After the colon is the expression of the anonymous function, which must fit on a single line. In fact, the definition of a lambda function is equivalent to:

```
def fun(parameter list):
    return expression
```

【例 4-20】匿名函数示例。

[Example 4-20] The anonymous function example.

```
f=lambda: 52
print(f())
t=lambda x: x**3
print(t(5))
add=lambda x, y: x+y
print(add(3.5, 4.7))
```

输出：

Output:

```
52
125
8.2
```

示例中，变量 f 所绑定的匿名函数不接收参数，函数返回值永远是 52；变量 t 所绑定的匿名函数接收一个参数，功能为求这个数的立方；变量 add 所绑定的匿名函数接收两个参数，功能为求这两个数的和。

In the example provided, the variable "f" is bound to an anonymous function that does not take any parameters and always returns 52. The variable "t" is bound to an anonymous function that takes one parameter and calculates its cube. The variable "add" is bound to an anonymous function that takes two parameters and calculates their sum.

4.6 递归函数
4.6 Recursive Functions

在计算机科学中递归是指一种通过重复将问题分解为同类的子问

In computer science, recursion is a method of solving problems by repeatedly breaking them down into

129

题来解决问题的方法。在 Python 中，一个函数直接或间接地调用自身，被称为函数的递归调用。递归特别适用于解决可以分解为更小、更简单的子问题的问题，比如计算阶乘、遍历树或图等。

递归调用通常能够将一个大型的复杂问题转化为规模较小的子问题，将复杂问题的递归条件，一层一层地回溯到终止条件，然后根据终止条件的运算结果，一层一层地递进运算到满足全部的递归条件。递归能用少量程序描述解决解题过程中的重复运算部分，减少代码量。但是使用递归调用时，要在编程时计划好终止条件，写好递归条件，用回溯的算法思想解决问题。

在 Python 中，递归一般是通过函数加分支结构来实现的，实现递归函数需要满足以下条件。

（1）必须有一个明确的递归终止条件。

（2）给出递归终止时的处理办法。

（3）每次进入更深一层递归时，问题规模相比上次递归都应有所减小或更接近解。

【例4-21】定义递归函数fun(n)，计算 n 以内的正整数累加和，并调用该函数求 100 以内的正整数累加和。

similar subproblems. In Python, when a function directly or indirectly calls itself, it is known as a recursive call of the function. Recursion is particularly useful for solving problems that can be decomposed into smaller, simpler subproblems, such as calculating factorials, traversing trees or graphs.

Recursive calls typically transform a large, complex problem into smaller subproblems. By following the recursive conditions of the complex problem, one can trace back through each layer to a termination condition. Then based on the computation results of the termination condition, one can progressively compute each layer to satisfy all the recursive conditions. Recursion can be used to describe the repetitive parts of the problem-solving process with minimal programming, reducing the amount of code. However, when using recursive calls, it's essential to plan termination conditions during programming, define recursive conditions, and solve problems using a backtracking algorithmic approach.

In Python, recursion is typically implemented using functions combined with conditional structures. To implement a recursive function in Python, the following conditions need to be satisfied.

(1)There must be a clear termination condition for recursion.

(2)Provide a solution when recursion terminates.

(3)Each time the recursion goes deeper, the size of the problem should decrease compared to the previous recursion or move closer to the solution.

[Example 4-21] Define a recursive function fun(n) to calculate the sum of positive integers within n, and call this function to find the sum of positive integers within 100.

```
def fun(n):
    if n==1:
        return 1
    else:
        return fun(n-1)+n
```

4-21

```
print(fun(100))
```

输出： | Output:

```
5050
```

【例4-22】定义递归函数fact(n)，计算 n 的阶乘，并调用该函数求 10 的阶乘。

[Example 4-22] Define a recursive function fact(n) to calculate the factorial of n, and call this function to find the factorial of 10.

```
def fact(n):
    if n==0:
        return 1
    return n*fact(n-1)

print(fact(10))
```

4-22

输出： | Output:

```
3628800
```

【例4-23】定义递归函数fib(n)，输出斐波那契数列的前 n 项，并使用该函数输出斐波那契数列的前 10 项。

[Example 4-23] Define a recursive function fib(n) to display the first n terms of the Fibonacci sequence, and use this function to output the first 10 terms of the Fibonacci sequence.

```
def fib(n):
    if n==0 or n==1:
        return 1
    else:
        return fib(n-2)+fib(n-1)

for i in range(10):
    print(fib(i),end=' ')
```

4-23

输出： | Output:

```
1 1 2 3 5 8 13 21 34 55
```

4.7 内置函数
4.7 Built-in Functions

在 Python 中，可以根据需求定义函数，但其实还有很多内置函

In Python, one can define functions based on needs. In fact, there are many built-in functions available for di-

数可以直接使用，如表 4-1 所示，具体语法可以参考官方帮助文档。

表 4-1 中的函数如果进行分类的话，大概可以分成 4 类：数学运算函数、字符串处理函数、转换函数和序列操作函数。常用的数学运算函数和字符串处理函数在前面章节已经介绍过，这里不再赘述。

rect use. Commonly used functions are listed in Table 4-1, and their specific syntax can be found in the official help documentation.

If the functions in Table 4-1 are classified, they can be roughly divided into four categories: mathematical calculation functions, string processing functions, conversion functions, and sequence operation functions. Commonly used mathematical calculation functions and string processing functions have already been introduced in previous chapters, and will not be elaborated here.

表 4-1 Python 内置函数
Table 4-1 Python Built-in Functions

abs()	chr()	exec()	hex()	map()	print()	str()
all()	classmethod()	filter()	id()	max()	property()	sum()
any()	compile()	float()	input()	memoryview()	range()	super()
ascii()	complex()	format()	int()	min()	repr()	tuple()
bin()	delattr()	frozenset()	isinstance()	next()	reversed()	type()
bool()	dict()	getattr()	issubclass()	object()	round()	vars()
breakpoint()	dir()	globals()	iter()	oct()	set()	zip()
bytearray()	divmod()	hasattr()	len()	open()	setattr()	
bytes()	enumerate()	hash()	list()	ord()	slice()	
callable()	eval()	help()	locals()	pow()	sorted()	

4.7.1 转换函数
4.7.1 Conversion Functions

转换函数主要用于不同数据类型之间的转换，常见的内置转换函数如表 4-2 所示。

Conversion functions are primarily used for converting between different data types. The common built-in conversion functions are shown in Table 4-2.

表 4-2 常见的内置转换函数
Table 4-2 Common Built-in Conversion Functions

函数名 Function Name	函数描述	Function Description
int(x)	从一个数字或字符串 x 构造一个整数对象	Return an integer object constructed from a number or string x

续表

函数名 Function Name	函数描述	Function Description
float(x)	从一个数字或字符串 x 构造一个浮点数	Return a floating-point number constructed from a number or string x
complex(x)	返回一个复数,其值为 real+imag *1j,或将字符串或数字转换为复数	Return a complex number with the value real+imag *1j or convert a string or number to a complex number
bool(x)	返回一个布尔值,即 True 或 False 之一。x 使用标准的真值测试程序进行转换。如果 x 为假或被省略,则返回 False；否则返回 True	Return a Boolean value, i.e. one of True or False. x is converted using the standard truth testing procedure. If x is false or omitted, it returns False; otherwise it returns True
str(object)	返回对象的字符串版本。如果未提供对象,则返回空字符串	Return a string version of object. If object is not provided, return the empty string
ord(c)	给定一个字符 c,返回一个整数,表示该字符的 Unicode 编码	Given a character c, return an integer representing the Unicode code of that character
chr(i)	返回一个 Unicode 编码 i 对应的字符。这是 ord() 的反操作	Return a character corresponding to a Unicode code i. This is the inverse of ord()
bin(n)	将一个整数 n 转换为以 0b 开头的二进制字符串	Convert an integer number n to a binary string prefixed with 0b
oct(n)	将一个整数 n 转换为以 0o 开头的八进制字符串	Convert an integer numbern to an octal string prefixed with 0o
hex(n)	将一个整数转换为以 0x 开头的十六进制字符串	Convert an integer number n to a hexadecimal string prefixed with 0x

```
print(ord('a'),ord('好'))
print(chr(97),chr(22909))
print(bin(10))
print(oct(10))
print(hex(10))
```

输出: Output:

```
97 22909
a 好
0b1010
0o12
0xa
```

4.7.2 序列操作函数
4.7.2 Sequence Operation Functions

序列是 Python 中一种重要的数据结构，包括字符串、列表、元组和 range 对象，表 4-3 中列出了常见的序列操作函数。

Sequence is an important data structure in Python, including strings, lists, tuples, and range objects. Table 4-3 lists commonly sequence for operations functious.

表 4-3　常见的序列操作函数
Table 4-3　Common Sequence Operation Functions

函数名 Function Name	函数描述	Function Description
range(start, stop[, step])	从 start（包括）到 stop（不包括）按步长生成一个整数序列并返回	Return an object that produces a sequence of integers from start (inclusive) to stop (exclusive) by step
sorted(iterable)	从可迭代对象 iterable 中返回一个新的排序列表	Return a new sorted list from the items in iterable
reversed(seq)	返回一个反向迭代器。seq 必须是一个具有 _reversed_() 方法的对象，或者支持序列协议（即具有 _len_() 方法和可以接收从 0 开始的整数参数的 _getitem_() 方法）	Return a reverse iterator. seq must be an object which has a _reversed_() method or supports the sequence protocol (the _len_() method and the _getitem_() method with integer arguments starting at 0)
map(func, * iterable)	返回一个迭代器，该迭代器将函数应用于可迭代对象的每个项目并产生结果。如果传入了额外的可迭代参数，则函数必须接收相同数量的参数，并且会并行应用于所有可迭代对象中的项。当存在多个可迭代对象时，迭代器在最短的可迭代对象耗尽时停止	Return an iterator that applies function to every item of iterable, yielding the results. If additional iterable arguments are passed, function must take that many arguments and is applied to the items from all iterables in parallel. With multiple iterables, the iterator stops when the shortest iterable is exhausted
zip(* iterables)	返回一个元组迭代器，其中，第 i 个元组包含每个参数序列或可迭代对象中的第 i 个元素。当最短的输入可迭代对象耗尽时，迭代器停止。如果只有一个可迭代参数，那么它将返回一个 1 元组的迭代器。如果没有参数，那么它将返回一个空迭代器	Return an iterator of tuples, where the i-th tuple contains the i-th element from each of the argument sequences or iterables. The iterator stops when the shortest input iterable is exhausted. With a single iterable argument, it returns an iterator of 1-tuple. With no arguments, it returns an empty iterator
filter(iterable)	从可迭代对象 iterable 中构造一个迭代器，包含那些函数值返回 True 的元素。可迭代对象可以是一个序列、一个支持迭代的容器或一个迭代器	Construct an iterator from those elements of iterable for which function returns True. Iterable may be either a sequence, a container which supports iteration, or an iterator

续表

函数名 Function Name	函数描述	Function Description
all(iterable)	如果可迭代对象 iterable 中的所有元素为 True(或者可迭代对象为空)，则返回 True	Return True if all elements of the iterable are True (or the iterable is empty)
any(iterable)	如果可迭代对象 iterable 中的任意元素为 True，则返回 True；如果可迭代对象为空，则返回 False	Return True if any element of the iterable is True. If the iterable is empty, return False

```
ls1=[4,3,2,5,1,0]
print(ls1)
ls2=sorted(ls1)
print(ls2)
ls3=list(reversed(ls2))
print(ls3)
str1=list(map(str,ls1))
print(str1)
ls4=list(zip(ls1,ls2))
print(ls4)
ls5=list(filter(lambda x : x % 2==1,ls1))
print(ls5)
print(all(ls1))
print(any(ls1))
```

输出： | Output:

```
[4, 3, 2, 5, 1, 0]
[0, 1, 2, 3, 4, 5]
[5, 4, 3, 2, 1, 0]
['4', '3', '2', '5', '1', '0']
[(4, 0),(3, 1),(2, 2),(5, 3),(1, 4),(0, 5)]
[3, 5, 1]
False
True
```

4.7.3 help()函数
4.7.3 help() Function

如果想了解某个函数的说明，可以使用 help()函数查看帮助信息。help()函数的语法格式为：

If one wants to understand the documentation of a function, one can use the help() function to view the help information. The syntax format of the help() function is:

help([object])

该函数可以返回方括号中对象的帮助信息，括号中的参数为字符串，可以是模块名、函数名、类名、方法名、关键字或文档主题。

当参数省略时，则进入帮助环境，输入查找的对象名，返回该对象的帮助信息。

例如，查看 print()函数的帮助信息。

This function can return help information for the object specified within the parentheses. The parameter within the parentheses is a string, which can be a module name, function name, class name, method name, keyword, or documentation topic.

When the parameter is omitted, it enters the help environment, and one can enter the name of the object to look up to return the help information for that object.

For example, to view the help information for the print() function.

```
>>>help(print)
Help on built-in function print in modulebuiltins:

print(…)
    print(value, …, sep=' ', end='\n', file=sys. stdout, flush=False)
    
    Prints the values to a stream, or tosys. stdout by default.
    Optional keyword arguments:
    file:   a file-like object(stream); defaults to the currentsys. stdout.
    sep:    string inserted between values, default a space.
    end:    string appended after the last value, default a newline.
    flush:  whether to forcibly flush the stream.
```

4.8 本章小结
4.8 Summary of This Chapter

本章介绍了 Python 中函数相关基础知识，包括函数的定义、函数调用和返回值、函数的参数传递以及变量的作用域等，主要内容如下。

（1）普通函数的定义、调用以及返回值，定义使用关键字 def，返回值使用关键字 return。

（2）函数的形参和实参。函数定义时括号中的参数称为形参，调用时括号中的参数称为实参。

This chapter introduces the basic knowledge related to functions in Python, including function definitions, function call and return values, parameter passing, and variable scope. The main contents are as follows.

(1) Definition, call and return values of ordinary functions. Use the keyword def for definition, and the keyword return for return values.

(2)Formal parameters and actual parameters of functions. Parameters in parentheses when defining a function are called formal parameters, while parameters in paren-

（3）函数的参数传递，包括位置传递、关键字传递、默认值传递、可变长参数传递和解包裹传递5种方式。

（4）变量作用域，包括局部变量和全局变量。以函数为边界，函数外部定义的变量是全局变量，函数内部定义的变量是局部变量，形参被视为局部变量。

（5）匿名函数、递归函数和内置函数。

(3) Parameter passing in functions, including positional, keyword, default value, variable-length, and unpacking parameter passing.

(4) Variable scope, including local and global variables. Variables defined outside functions are global variables, while variables defined inside functions are local variables, and formal parameters are regarded as local variables.

(5) Anonymous functions, recursive functions, and built-in functions.

4.9 本章练习
4.9 Exercise for This Chapter

4.1 编写一个函数，打印出所有小于100的自然数中是3或5的倍数的数的总和。

4.2 编写一个函数，接收一个字符串作为参数，返回字符串中元音字母(aeiou)的数量。

4.3 给定一个正整数(输入值小于1 000)，编写程序计算小于或等于该数的素数的累加和。其中素数的判断由函数实现。

4.4 编写函数 change(str)，其功能是对参数 str 进行大小写转换，即字符串中的大写字母转换为小写字母，小写字母转换为大写字母。

4.5 编写函数 digit(num,k)，其功能是求整数 num 的第 k 位的值。

4.6 如果一个数恰好等于它的真因子(除自身以外的约数)之和，这个数就被称为完数。编写函数判断该数是否为完数，并编程找出1 000以内的所有完数。

4.1 Write a function to print the sum of all natural numbers less than 100 that are multiples of 3 or 5.

4.2 Write a function that takes a string as a parameter and returns the number of vowels(aeiou) in the string.

4.3 Given a positive integer(input value less than 1,000), write a program to calculate the cumulative sum of prime numbers less than or equal to that number. The determination of prime numbers is implemented by a function.

4.4 Write a function change(str) that converts the case of the parameter str, converting uppercase letters in the string to lowercase and lowercase letters to uppercase.

4.5 Write a function digit(num, k) that calculates the value of the k-th digit of the integer num.

4.6 If a number is called a perfect number, if it is equal to the sum of its proper divisors(excluding itself). Write a function to determine whether a number is a perfect number, and program to find all perfect numbers within 1,000.

4.7 求斐波那契数列（1，1，2，3，5，8，13，21…）的前 30 项并输出，并控制每行输出 5 个。

4.8 编写函数 reverse(x)，输入一个正整数，并将各位数字反转后输出。

4.9 编写一个函数，接收一个整数作为参数，将其转换为字符串，然后判断这个字符串是否为回文数（正读和反读都一样）。

4.7 Calculate the first 30 terms of the Fibonacci sequence(1, 1, 2, 3, 5, 8, 13, 21,…) and output them, controlling the output to 5 numbers per line.

4.8 Write a function reverse(x) that takes a positive integer as input and outputs the reversed digits.

4.9 Write a function that takes an integer as a parameter, converts it to a string, and then checks if the string is a palindrome(reads the same forward and backward).

本章练习题参考答案

第 5 章　列表与元组
Chapter 5　List and Tuple

Python 中的序列包含列表(list)、元组(tuple)、字符串(string)等，它们都可用来存储一系列有序元素。列表是可变的，可以随时添加、删除或修改其中的元素；而元组是不可变的，一旦创建就不能修改其内容。在语法上，列表使用方括号[]创建，元素之间用逗号分隔；元组则使用圆括号()创建。列表和元组都支持索引和切片操作，方便访问和处理集合中的元素。

Sequences in Python, including lists, tuples, strings, are used to store a collection of ordered elements. Lists are mutable, allowing elements to be added, removed, or modified at any time, while tuples are immutable, meaning their content cannot be changed once created. In terms of syntax, lists are created using square brackets [] with elements separated by commas, while tuples are created using parentheses (). Both lists and tuples support indexing and slicing operations, making it easy to access and manipulate elements within the collection.

5.1　列表
5.1　List

列表是序列类型的一种，能够存储多种类型的有序元素。它支持各种操作，如添加、删除、修改和查找元素。通过索引或切片操作可以访问列表中的某一个元素或某几个元素。列表的可变性意味着其内容可以根据需要动态调整。

A list is a type of sequence that can store ordered elements of various types. It supports operations such as adding, deleting, modifying, and searching for elements. Indexing or slicing can be used to access a specific element or a subset of elements in the list. The mutability of lists means that their content can be dynamically adjusted as needed.

列表举例：

Examples of lists:

[11, 22, 33, 40]
['red flower', 'blue sky', 'green leaf']
['Python', 3.14, 5, [10, 22]]
[['file1', 200, 7],'file2', 263, 9]

5.1.1 列表的创建
5.1.1 Creating Lists

Python 中的列表是一种可变的数据结构,它可以包含不同类型的元素,如整数、浮点数、字符串甚至其他列表(嵌套列表)。以下是 Python 中列表的几种创建方式。

List in Python is a mutable data structure that can contain elements of different types, such as integers, floating-point numbers, strings, and even other lists (nested lists). Here are several ways to create lists in Python.

(1)使用方括号创建空列表:这是创建空列表的最直接方式,之后可以向其中添加元素。

(1)Creating an empty list using square brackets: This is the most direct way to create an empty list, and elements can be added to it later.

```
my_list=[]
empty_list=[] #Create an empty list
```

(2)使用方括号创建带有初始元素的列表:在方括号中列出要加入列表的元素,用逗号分隔。这种方式可以在创建列表时直接初始化元素。

(2)Using square brackets to create a list with initial elements: List the elements to be added to the list in square brackets, separated by commas. This method allows initializing elements directly when creating the list.

```
my_list=[element1, element2, …, elementN]
```

例如:创建一个包含初始元素的列表。

For example: Creating a list with initial elements.

```
my_list=[1, 2, 3, 'apple', 'banana']
```

(3)使用 list()函数创建空列表:list()函数可以不带参数地调用以创建空列表,或者传递一个可迭代对象(如字符串、元组、集合等)作为参数,以创建包含该可迭代对象元素的列表。

(3)Creating an empty list using the list() function: The list() function can be called without any arguments to create an empty list, or it can be passed an iterable object (such as a string, tuple, set, etc.) as an argument to create a list containing elements of that iterable object.

```
my_list=list() or my_list=list(iterable)
```

例如:从其他可迭代对象创建列表。

For example: Creating a list from other iterable objects.

```
chars_list=list('hello')
#This will create a list containing each character from the string 'hello'.
```

5.1.2 元素的索引及访问
5.1.2 Indexing and Accessing Elements

在 Python 中，列表是一种可变的数据结构，它包含了一组有序的元素，这些元素可以是不同的类型（例如整数、浮点数、字符串等），并且可以通过索引进行访问。索引是列表中元素的位置或编号，正向索引从 0 开始到列表长度减 1 结束，负向索引从 -1 开始到列表长度取负值结束。

In Python, a list is a mutable data structure that contains an ordered collection of elements, which can be of different types(such as integers, floats, strings, etc.), and can be accessed by index. An index is the position or number of an element in the list, with positive indexing starting from 0 and going up to the length of the list minus 1, negative indexing starts from −1 and goes down to the negative length of the list.

（1）正向索引是从 0 开始的整数，表示列表中的位置。第一个元素的索引是 0，第二个元素的索引是 1，依次类推。如果尝试访问超出列表长度的正向索引，Python 将会引发 IndexError 异常。

(1) Positive indexing starts from integer 0, representing the positions in the list. The index of the first element is 0, the second element is 1, and so on. If attempting to access a positive index beyond the length of the list, Python will raise an IndexError exception.

【例 5-1】正向索引示例。

[Example 5-1] Example of positive indexing.

```
my_list=['apple', 'banana', 'cherry', 'date']
#Accessing the first element(index 0)
first_element=my_list[0]
print(first_element) #'apple'
#Accessing the fourth element(index 3)
third_element=my_list[3]
print(third_element) #'date'
#Accessing the second element(index 1)
my_list[1]='berry'
print(my_list) #my_list now is ['apple', 'berry', 'cherry', 'date']
```

5-1

（2）负向索引是从 -1 开始的整数，表示从列表的末尾开始反向计数，编码方向为从右至左。最后一个元素的索引是 -1，倒数第二个元素的索引是 -2，依次类推。

(2) Negative indexing starts from integer −1, representing reverse counting from the end of the list, with the indexing direction from right to left. The index of the last element is −1, the second-to-last element is −2, and so forth.

【例 5-2】负向索引示例。

[Example 5-2] Example of negative indexing.

141

```
my_list=['apple', 'banana', 'cherry', 'date']
#Accessing the last element(index -1)
last_element=my_list[-1]
print(last_element) #'date'
#Accessing the second-to-last element(index -2)
second_last_element=my_list[-2]
print(second_last_element) #'cherry'
#Modifying the third-to-last element(index -3)
my_list[-3]='strawberry'
print(my_list) #my_list now is['apple', 'strawberry', 'cherry', 'date']
```

使用负向索引时和正向索引一样，如果尝试访问超出列表范围的负向索引（比如 my_list[-5] 对于只有 4 个元素的列表来说），Python 同样将会引发 IndexError 异常。

正向索引和负向索引都是访问和修改 Python 列表中元素的有效方式。正向索引从列表的开头开始正向计数，而负向索引从列表的末尾开始反向计数。在实际编程中，根据具体情况选择使用正向或负向索引可以使代码更加简洁和易读。

When using negative indexing, similar to positive indexing, if an attempt is made to access a negative index that is beyond the range of the list(for example, my_list[-5] for a list with only has 4 elements), Python will also raise an IndexError exception.

Both positive and negative indexing are effective ways to access and modify elements in a Python list. Positive indexing counts forward from the beginning of the list, while negative indexing counts from the end of the list in reverse. In practical programming, choosing to use positive or negative indexing based on the specific situation can make the code more concise and readable.

5.1.3 切片操作
5.1.3 Slicing Operation

在 Python 中，列表的切片操作是一种非常灵活的方式，用于获取列表的子集。切片操作使用 3 个参数：起始索引、结束索引和步长。3 个参数都必须为整数类型。基本语法如下：

In Python, slicing operation on a list is a very flexible way to obtain a subset of the list. The slicing operation uses three parameters: start index, end index, and step size, all of which must be integers. The basic syntax is as follows:

```
list[start:end:step]
```

（1）start：起始索引，表示切片开始的位置。如果省略，则默认为 0。

（2）end：结束索引，表示切

(1)start: The starting index that indicates where the slicing starts. If omitted, it defaults to 0.

(2)end: The ending index that indicates where the

第5章 列表与元组
Chapter 5 List and Tuple

片结束的位置(但不包括该位置的元素)。如果省略,则切片将一直进行到列表的末尾。

(3)step:步长,表示切片中每个元素之间的间隔,步长可以为正整数也可以为负整数。如果省略,则默认为1。

【例5-3】列表切片示例。

slicing ends(but does not include the element at that position). If omitted, the slicing will continue to the end of the list.

(3) step: The step size, indicating the interval between each element in the slice, which can be a positive or negative integer. If omitted, it defaults to 1.

[Example 5-3] Example of list slicing.

```
#Define a list
my_list=[0, 1, 2, 3, 4, 5, 6, 7, 8, 9]
#Get the elements between index 2 and 6(excluding 6)
slice1=my_list[2:6]
print(slice1) #Output:[2, 3, 4, 5]
#Get all elements starting from index 4 to the end of the list
slice2=my_list[4:]
print(slice2) #Output:[4, 5, 6, 7, 8, 9]
#Get all elements from the beginning of the list up to index 6(excluding 6)
slice3=my_list[:6]
print(slice3) #Output:[0, 1, 2, 3, 4, 5]
#Get all elements from the beginning to the end of the list, but with a step size of 2
slice4=my_list[::2]
print(slice4) #Output:[0, 2, 4, 6, 8]
#Get all elements from the end to the beginning of the list, with a step size of -1(i.e., reverse the list)
slice5=my_list[::-1]
print(slice5) #Output:[9, 8, 7, 6, 5, 4, 3, 2, 1, 0]
```

当step为正整数时,start和end两个参数的设定需要遵循一定的规则,切片方向需要是从左向右,否则会返回一个空列表。例如:

When the step is a positive integer, the settings of the start and end parameters need to follow certain rules, and the slicing direction should be from left to right, otherwise, it will return an empty list. For example:

```
my_list=[1, 2, 3, 4, 5]
print(my_list[1:-3])    #Output:[2]
print(my_list[-3:-1])   #Output:[3,4]
print(my_list[-2:-4])   #Output:[]
```

当step为负整数时,将列表中的元素进行倒序输出,此时注意start和end参数的设置,切片方向需要是从右向左,否则会返回一个空列表。例如:

When the step is a negative integer, the elements in the list will be output in reverse order. At this point, pay attention to the setting of the start and end parameters, the slicing direction should be from right to left; otherwise, it will return an empty list. For example:

143

```
my_list=[1, 2, 3, 4, 5]
print(my_list[-1:-3])        #Output:[]
print(my_list[-1:-3:-1])     #Output:[5, 4]
print(my_list[-2:1:-1])      #Output:[4, 3]
```

5.2 操作列表元素
5.2 Manipulate List Elements

在 Python 中，列表具有可变性，这意味着不仅可以更改元素的值，还可以动态地添加或删除元素，从而改变列表的大小和结构。

In Python, lists are mutable, which means that one cannot only change the values of elements but also dynamically add or remove elements, thus altering the size and structure of the list.

5.2.1 增加列表元素
5.2.1 Adding Elements to a List

（1）append()方法：在列表末尾添加一个元素。这个方法会改变原来的列表，而不是创建一个新的列表。基本语法如下：

(1) append() method: Add an element to the end of the list. This method modifies the original list directly instead of creating a new one. The basic syntax is as follows:

```
list.append(obj)
```

obj 是想要添加到列表末尾的对象。例如：

obj is the object one wants to add to the end of the list. For example:

```
#Create an empty list
my_list=[]
#Add elements using the append() method
my_list.append("apple")
my_list.append("banana")
my_list.append("cherry")
#Print the list
print(my_list)    #Output:['apple', 'banana', 'cherry']
```

在这个例子中，首先创建了一个空列表 my_list，然后使用 append() 方法添加了 3 个字符串元素。最后，打印出列表，可以看到它现在包含这 3 个元素。

In this example, first create an empty list my_list, then add three string elements using the append() method. Finally, print the list to see that it now contains these three elements.

（2）insert()方法：在指定索引

(2) insert() method: Insert an element at a specified

位置插入一个元素。

与 append() 方法不同，insert() 允许选择插入新元素的位置，而不仅仅是列表的末尾。基本语法如下：

Unlike the append() method, insert() allows one to choose the position for inserting a new element, not just at the end of the list. The basic syntax is as follows:

```
list.insert(index, obj)
```

index 是想要插入新元素的位置的索引。注意，列表的索引是从 0 开始的。如果 index 超出了列表的范围，则 Python 将会引发 IndexError 异常。obj 是想要插入列表中的对象。例如：

index is the index of the position where one wants to insert the new element. Note that the index of the list starts from 0. If the index is out of the list's range, Python will raise IndexError exception. obj is the object one wants to insert into the list. For example:

```
#Create a list
fruits = ['apple', 'banana', 'cherry']
#Insert 'orange' at index 1 using the insert() method
fruits.insert(1, 'orange')
#Print the list
print(fruits)   #Output:['apple', 'orange', 'banana', 'cherry']
```

在这个例子中，fruits 列表中包含 3 个字符串元素。使用 insert() 方法在索引 1 的位置插入了'orange'。因此，'orange'被插入'apple'和'banana'之间。最终，打印出修改后的列表。

In this example, the fruits list contains three string elements. 'orange' is inserted at index 1 using the insert() method. Therefore, 'orange' is inserted between 'apple' and 'banana'. Finally, print the modified list.

insert()方法会改变原来的列表，因为它是修改列表，而不是创建一个新的列表。如果 index 等于列表的长度（即 len(list)），那么 insert() 的执行操作将与 append() 相同，新元素将被添加到列表的末尾。

It is important to note that the insert() method will change the original list because it modifies the list instead of creating a new list. If the index is equal to the length of the list(i. e. len(list)), the behavior of insert() will be the same as append(), as the new element will be added to the end of the list.

5.2.2 删除列表元素
5.2.2 Delete List Elements

（1）remove()方法：删除列表中第一个出现的指定值。如果列表中没有该值，则 Python 将会引发

(1)remove() method: Remove the first occurrence of the specified value in the list. If the value is not found in the list, Python raises ValueError exception. The basic

ValueError 异常。基本语法如下： | syntax is as follows:

```
#Create a list
fruits=['apple', 'banana', 'cherry', 'banana', 'orange']
#Use the remove() method to remove the first 'banana'
fruits.remove('banana')
#Output the modified list
print(fruits)   #Output: ['apple', 'cherry', 'banana', 'orange']
```

虽然 fruits 列表中包含两个相同的字符串元素'banana'，但是 remove() 方法只删除列表中第一个与'banana'匹配的元素。

Since the list fruits contains two identical string elements 'banana', the remove() method will only delete the first matching element in the list.

（2）pop()方法：删除并返回指定索引处的元素，如果不提供索引，则默认删除并返回最后一个元素。当使用 pop()方法时，列表中的元素数量会减少，并且被删除的元素会作为返回值。基本语法如下：

(2)pop() method: Removes and returns the element at the specified index, or defaults to remove and return the last element if no index is provided. When using the pop() method, the number of elements in the list decreases, and the deleted element is returned as the result. The basic syntax is as follows:

```
element=list.pop([index])
```

element 是被删除元素的值，index 是可选参数，表示想要删除的元素的索引。如果未指定索引，则默认为-1，即删除并返回最后一个元素。

element is the value of the element being removed, index is an optional parameter that indicates the index of the element to remove. If no index is specified, the default is -1, which removes and returns the last element.

【例 5-4】pop()方法示例。

[Example 5-4] Example of pop() method.

```
my_list=[1, 2, 3, 4, 5]
#Remove and return the last element
last_element=my_list.pop()
print(last_element)           #Output:5
print(my_list)                #Output:[1, 2, 3, 4]
#Remove and return the element at index 1
element_at_index_1=my_list.pop(1)
print(element_at_index_1)   #Output:2
print(my_list)                #Output:[1, 3, 4]
```

5-4

remove()和pop()方法在列表中找不到指定元素时会引发 ValueError 异常，因此在使用这些方法之前，需要检查元素是否存在于列表中。

使用索引直接删除或修改元素时，如果索引超出列表范围，则 Python 会引将发 IndexError 异常。

（3）del 语句：用于从内存中删除对象，或者从数据结构（如列表、字典或集合）中移除元素。del 语句的语法很简单，但使用它时需要小心，因为一旦对象被删除，将无法再访问它，除非有其他引用指向同一个对象。

The remove() and pop() methods will raise a ValueError if the specified element is not found in the list, so it is necessary to check if the element exists in the list before using these methods.

When deleting or modifying elements directly using an index, Python will raise IndexError if the index is out of the range of the list.

(3)The del statement: Used to delete objects from memory or remove elements from data structures such as lists, dictionaries, or sets. The syntax of the del statement is straightforward, but caution is needed when using it because once an object is deleted, it cannot be accessed anymore unless there are other references pointing to the same object.

```
my_list=[1, 2, 3, 4]
del my_list[1]          #Delete the element at index 1
print(my_list )         #Output:[1,3,4]
del my_list[1:3]        #Delete the elements from index 1 up to 3(excluding 3)
print(my_list )         #Output:[1]
```

```
my_list=[1,2,3]
del my_list             #Delete 'my_list'
print(my_list)          #Output the exception information: NameError: name 'my_list' is not defined
```

尝试删除不存在的变量、列表元素或字典键时会引发错误（如 NameError 或 KeyError）。在删除列表元素时，要注意列表的大小会改变，这可能会影响迭代过程中的索引。

（4）clear()方法：删除列表中的所有元素。clear()是一个常见的方法，主要用于清空一些数据结构中的内容。

Attempt to delete a non-existent variable, list element, or dictionary key will raise an error(such as NameError or KeyError). When deleting list elements, it is important to note that the size of the list will change, which may affect the indexing during iteration.

(4)clear() method: Remove all elements from the list. clear() is a common method used to empty the contents of certain data structures.

```
my_list=[1, 2, 3]
my_list.clear()         #'my_list' is now an empty list []
print(my_list)          #Output:[]
```

5.2.3 修改列表元素
5.2.3 Modify List Elements

在 Python 中，修改列表中的元素非常简单和直接。由于列表是可变的，所以可以直接通过索引来访问并修改列表中的任何元素。以下是修改列表中元素的基本方法。

In Python, modifying elements in a list is very simple and straightforward. Since lists are mutable, one can directly access and modify any element in the list using its index. Here are the basic methods for modifying elements in a list.

(1)通过索引修改：可以使用索引直接访问列表中的元素，并为其分配新的值。

(1)Modifying by index: One can use an index to directly access an element in a list and assign it a new value.

```
my_list=[1, 2, 3, 4, 5]
#Modify the element at index 2(i.e., the third element)
my_list[2]=10
print(my_list)   #Output:[1, 2, 10, 4, 5]
```

(2)通过切片修改：可以使用切片来修改列表中多个元素的值。例如，替换 my_list 列表中索引 1 到 4(不包括 4)的元素。

(2) Modifying by slicing: One can use slicing to modify the values of multiple elements in a list. For example, replace the elements at indexes 1 to 4(excluding 4) in the my_list list.

```
my_list=[1, 2, 3, 4, 5]
#Use slicing to modify multiple elements in a list
my_list[1:4]=[20, 30, 40]
#Output the modified list
print(my_list)   #Output:[1, 20, 30, 40, 5]
```

(3)使用循环修改：可以使用循环来遍历列表，并根据某些条件修改元素。

(3)Modifying with a loop: One can use a loop to iterate through a list and modify elements based on certain conditions.

```
my_list=[1, 2, 3, 4, 5]
#Modify each element in the list using a loop
for i in range(len(my_list)):
    my_list[i]*=2   #Multiply each element by 2
#Output the modified list
print(my_list)   #Output:[2, 4, 6, 8, 10]
```

5.2.4 列表排序操作
5.2.4 List Sorting Operation

（1）sort()：对列表中的元素进行排序，默认为升序排列。基本语法如下：

(1)sort(): Sort the elements in a list, by default in ascending order. Basic syntax is as follows:

```
list.sort(key=None, reverse=False)
```

key 参数：为进行比较的元素指定以何种形式进行排序，例如，key=len 会根据列表中每一个元素的长度进行排序。如果不指定 key 参数，则默认使用元素本身的比较规则。此参数可省略。

key parameter: This is used to specify the form of comparison for sorting, for example, key=len will sort based on the length of each element in the list. If the key parameter is not specified, the default comparison rule of the elements will be used. This parameter is optional.

reverse 参数：用于指定排序规则，reverse=True 表示降序，reverse=False 表示升序（默认），此参数可省略。

reverse parameter: This is used to specify the sorting rule where reverse=True specifies descending order, and reverse=False specifies ascending order(by default). This parameter is optional.

【例 5-5】sort()方法示例。

[Example 5-5] Example of sort() method.

```
#Default ascending order sorting
numbers=[5, 1, 9, 3, 7]
numbers.sort()
print(numbers)   #Output: [1, 3, 5, 7, 9]
#Descending order sorting
numbers.sort(reverse=True)
print(numbers)   #Output: [9, 7, 5, 3, 1]

#Sorting using the key parameter
words=['bananas', 'cherries', 'apples']
words.sort(key=len)   #Sort by string length
print(words)   #Output: ['apples', 'bananas', 'cherries']
```

（2）sorted()：对列表中的元素进行排序。基本语法如下：

(2)sorted(): Sort the elements in a list. Basic syntax is as follows:

```
sorted(iterable, key=None, reverse=False)
```

iterable 参数：可迭代对象。

iterable parameter: Iterable object.

key 参数：使用同 sort()方法中的 key 参数，在此不再赘述。

key parameter: The key parameter functions the same as in the sort() method, and it is not further elaborated here.

reverse 参数：排序规则。reverse = True 表示降序，reverse = False 表示升序（默认）。

reverse parameter: Sorting rule where reverse = True specifies descending order, and reverse = False specifies ascending order(default).

```
#Default ascending order sorting
numbers = [5, 1, 9, 3, 7]
sorted_numbers = sorted(numbers)
print(sorted_numbers)            #Output: [1, 3, 5, 7, 9]
print(numbers)                   #Output: [5, 1, 9, 3, 7] the original list remains unchanged

#Descending order sorting
sorted_numbers_desc = sorted(numbers, reverse = True)
print(sorted_numbers_desc)       #Output: [9, 7, 5, 3, 1]

#Sorting using the key parameter
words = ['apples', 'cherries', 'bananas']
sorted_words = sorted(words, key = len)    #Sort by string length
print(sorted_words)              #Output: ['apples', 'bananas', 'cherries']
```

在 Python 中，sorted()和 sort()都是用于排序的函数/方法，但它们之间有一些关键的区别。

sorted()是 Python 的内置函数。它接收一个可迭代对象（如列表、元组等）作为输入，并返回一个新的已排序列表，原列表不会被修改。sorted()可以使用额外的参数来自定义排序方式，比如 key 和 reverse。

sort()是列表对象的一个方法。它没有返回值（或者说返回 None），而是直接对原列表进行排序，修改原列表。它同样可以使用 key 参数和 reverse 参数来自定义排序方式。

在实际应用中，如果需要保留原列表不变，并获取一个已排序的副本，那么使用 sorted()；如果不需要保留原列表，并希望直接修改它，那么使用 sort()。

（3）reverse()：将列表中的元素逆序排列。reverse()是列表对象

In Python, both sorted() and sort() are functions/methods used for sorting, but they have some key differences.

sorted() is a built-in function in Python. It takes an iterable object(like a list, tuple, etc.) as input and returns a new sorted list. The original list is not modified. Additional parameters can be used to customize the sorting, such as key and reverse.

sort() is a method of list objects. It does not return a value(or returns None) but directly sorts the original list to modify it. Similarly, key parameter and reverse parameter can be used to customize the sorting.

In practical applications, if there is a need to preserve the original list and obtain a sorted copy, then sorted() should be used; if there is no need to preserve the original list and prefer to modify it directly, then sort() should be used.

(3)reverse(): Reverse the elements in the list. reverse() is a method of list objects, used to reverse the elements in

的一个方法，用于就地将列表中的元素逆序排列。它没有参数，并且不返回任何值（实际上是返回None）。

the list in-place. It takes no arguments and does not return any value(actually returns None).

```
my_list=[1, 2, 3, 4, 5]
my_list.reverse()      #Reverse the order of elements in the list
print(my_list)         #Output: [5, 4, 3, 2, 1]
```

reverse()只能用于列表，并且会直接修改原列表。sort()和reverse()方法都会修改列表本身，而不是返回一个新的列表。如果想要保留原始列表并创建一个排序或逆序后的新列表，可以使用切片运算符来复制列表，然后对复制的列表调用这些方法，或者使用内置函数sorted()来获取排序后的新列表。

reverse() can only be used with lists and directly modifies the original list. Both sort() and reverse() methods will modify the list itself instead of returning a new list. If one wants to preserve the original list and create a sorted or reversed new list, one can use the slice operator to copy the list and then call these methods on the copied list, or use the built-in function sorted() to get a new sorted list.

（4）reversed()：是Python的内置函数，它接收一个序列作为参数，并返回一个逆序的迭代器。这个迭代器可以遍历原始序列中的元素，元素是逆序排列的。

(4)reversed(): Is a built-in function in Python that takes a sequence as an argument and returns a reverse iterator. This iterator can iterate over the elements of the original sequence in reverse order.

```
my_list=[1, 2, 3, 4, 5]
reversed_list=reversed(my_list)   #Return a reversed iterator
print(list(reversed_list))         #Output: [5, 4, 3, 2, 1]
```

reversed()可以用于任何序列类型（如列表、元组、字符串等），但它返回的是一个迭代器，而不是一个新的列表或其他具体的数据结构。如果需要一个完整的反转列表或其他数据结构，则需要将迭代器的内容转换回相应的类型，如上例中使用list()函数将迭代器转换为列表。

The reversed ( ) function can be used on any sequence type(such as lists, tuples, strings, etc.), but it returns an iterator rather than a new list or other specific data structure. If a complete reversed list or other data structure is needed, the contents of the iterator should be converted back to the corresponding type, like using the list() function as shown in the example above.

reverse()方法和reversed()方法的主要区别是：reverse()是列表的一个方法，无参数，直接修改原列表；reversed()是Python的内置函

The main difference between the reverse() method and the reversed() method is that reverse() is a method of lists, takes no arguments, and directly modifies the original list. On the other hand, reversed() is a built-in

数，接收一个序列作为参数，并返回一个反转的迭代器。

function in Python, takes a sequence as an argument, and returns a reversed iterator.

5.2.5 列表推导式
5.2.5 List Comprehension

列表推导式是 Python 中一种非常强大且简洁的构造列表的方法。它可以通过一行代码生成一个新的列表，该列表是通过迭代一个已有的可迭代对象（如列表、元组、字符串等）并对每个元素使用一定的表达式或条件得到的。列表推导式的基本语法如下：

List comprehension is a powerful and concise way to construct lists in Python. It allows one to generate a new list in a single line of code by iterating over an existing iterable object(such as a list, tuple, string, etc.) and applying a certain expression or condition to each element. The basic syntax of list comprehension is as follows:

```
[expression for item in iterable if condition]
```

expression：这是想要添加到新列表中的值，它是基于 item 的任何表达式或函数调用。

for item in iterable：迭代部分，item 是从 iterable 中每次迭代获取的元素。

if condition（可选）：条件语句，只有当 condition 为 True 时，expression 的结果才会被添加到新列表中。

expression: This is the value to be added to the new list, based on any expression or function call involving the item.

for item in iterable: This is the iteration part, where item is the element obtained from iterable in each iteration.

if condition(optional): This is a conditional statement where the result of the expression is added to the new list only if the condition is True.

【例5-6】列表推导式示例。

[Example 5-6] Example of list comprehension.

```
#Create a list of even numbers from 0 to 9
even_numbers=[x for x in range(10) if x % 2==0]
print(even_numbers)        #Output: [0, 2, 4, 6, 8]

#Create a list containing the ASCII values of each character in a string.
ascii_values=[ord(c) for c in 'hello']
print(ascii_values)        #Output: [104, 101, 108, 108, 111]

#Create a list containing the squares of each number in the list
squares=[x**2 for x in [1, 2, 3, 4, 5]]
print(squares)        #Output: [1, 4, 9, 16, 25]
```

5-6

列表推导式不仅使代码更加简洁，而且通常比使用传统的 for 循环和 append() 结合添加元素更快。

List comprehensions not only make the code more concise but also are often faster than using traditional for loops and appending elements. However, for complex

然而，对于复杂的逻辑或需要多行代码来处理每个元素的情况，使用传统的 for 循环可能更加清晰和易于理解。

logic or situations where multiple lines of code are required to process each element, using traditional for loops may be clearer and easier to understand.

5.3 元组
5.3 Tuple

元组是 Python 中的一种数据类型，它是一个不可变的序列，即一旦创建就不能修改其内容。元组通常用于存储一组相关的值，这些值可以是不同类型的数据。

A tuple is a data type in Python that is immutable, meaning its content cannot be changed once created. Tuples are commonly used to store a group of related values, which can be of different data types.

元组的创建非常简单，只需要将一系列元素用逗号分隔，并放在圆括号中即可。

Creating a tuple is very simple, it just involves separating a series of elements by commas and enclosing them in parentheses.

5.3.1 元组的创建
5.3.1 Creating Tuples

元组和列表都属于序列类型中的一种，但是在表达方式上需要区分开。列表是使用方括号"[]"括起来的一组元素，元素之间用逗号","分隔；而元组是使用圆括号"()"将一组元素括起来，元素之间用逗号分隔。

Tuples and lists are both types of sequences, but they are distinguished by their syntax. Lists are a group of elements enclosed in square brackets "[]", with elements separated by commas ",". On the other hand, tuples are a group of elements enclosed in parentheses "()", also separated by commas.

元组是不可变的数据结构，它可以包含不同类型的元素，如整数、浮点数、字符串等。以下是元组的几种创建方式。

Tuples are immutable data structures that can contain elements of different types such as integers, floating-point numbers, strings, etc. Here are several ways to create tuples.

(1)使用圆括号创建空元组：这是创建空元组的最直接方式。语法：

(1)Create an empty tuple using parentheses: This is the most direct way to create an empty tuple. Syntax:

```
my_tuple=()
```

(2)使用圆括号创建带有初始元素的元组：在圆括号中列出要加入元组的元素，用逗号分隔。这种方式可以在创建元组时直接初始化元素。语法：

(2)Create a tuple with initial elements using parentheses: List the elements to be added to the tuple in parentheses, separated by commas. This method allows one to initialize the elements directly when creating the tuple. Syntax:

my_tuple=(element1, element2, …, elementN)

例如：

For example:

my_tuple=(1, 2, 3, 'apple', 'banana')

（3）使用 tuple()函数创建空元组或基于其他可迭代对象创建元组：tuple()函数可以不带参数地调用以创建空元组，或者传递一个可迭代对象（如字符串、列表、集合等）作为参数，以创建包含该可迭代对象元素的元组。语法：

(3) Using the tuple() function to create an empty tuple or create a tuple based on another iterable object: The tuple() function can be called without any arguments to create an empty tuple, or with an iterable object(such as a string, list, set, etc.) as an argument to create a tuple containing elements from that iterable object. Syntax:

my_tuple=tuple()

my_tuple=tuple(iterable)

例如，从其他可迭代对象创建元组：chars_tuple = tuple('hello')（这将创建一个包含字符串'hello'中每个字符的元组）。

For example, creating a tuple from anotheriterable object: chars_tuple=tuple('hello')(This will create a tuple containing each character from the string 'hello').

由于元组是不可变的，所以没有像列表那样的修改操作（如 append()方法或索引赋值）。任何尝试修改元组内容的操作都会导致 TypeError 异常。

Since tuples are immutable, they do not have modification operations like lists(such as the append() method or index assignment). Any attempt to modify the contents of a tuple will result in a TypeError exception.

5.3.2 序列通用操作
5.3.2 Common Sequence Operations

在 Python 中，序列包括字符串、列表和元组等类型。这些序列类型共享一些通用的操作和方法。以下是一些常见的序列操作。

In Python, sequences include types such as strings, lists, and tuples. These sequence types share some common operations and methods. Here are some common sequence operations.

（1）索引。序列中的每个元素都有两个索引编码（正向索引编码和逆向索引编码），可以使用索引来访问特定的元素。在 Python 中，索引值分为正向（从左至右从 0 开始，到序列长度减 1 结束）和负向（从右至左从 -1 开始，到序列长度取负值结束）。

(1)Indexing. Each element in a sequence is assigned two index codes (positive indexing and negative indexing), which can be used to access specific elements. In Python, index values are divided into positive(starting from 0 from the left to the sequence length minus 1) and negative(starting from -1 from the right to the negative sequence length).

```
my_list=[10, 20, 30, 40, 50]
print(my_list[2])    #Output: 30
```

（2）切片。切片操作允许访问序列的子集或全部，而不仅是单个元素。切片操作使用冒号"："来分隔起始索引和结束索引以及步长。

(2)Slicing. Slicing allows access to a subset or all of a sequence, rather than just individual elements. Slicing uses a colon ":" to separate the start index, end index, and step size.

```
my_tuple=(10, 20, 30, 40, 50)
print(my_tuple[1:4:2])      #Output:(20,40)
print(my_tuple[::-1])       #Output:(50, 40, 30, 20, 10)
```

（3）获取序列长度。len()函数可以用于获取序列的长度。

(3) Obtain the length of a sequence. The len() function can be used to obtain the length of a sequence.

```
my_tuple=(10, 20, 30, 40, 50)
print(len(my_tuple))   #Output: 5
```

（4）count()。统计序列中某个元素的出现次数。

(4)count(). To count the occurrences of a specific element in a sequence.

```
tuple1=(1, 2, 3, 2)
count_tuple1=tuple1.count(2)
print(count_tuple1) #Output: 2
```

（5）max()。max()方法是 Python 中的一个内置函数，用于返回可迭代对象（如列表、元组等）中的最大值，或者返回多个参数中的最大值。语法：

(5)max(). The max() method is a built-in function in Python used to return the maximum value in an iterable object(such as list, tuple, etc.), or return the maximum value among multiple arguments. Syntax:

```
max(iterable)
max(arg1, arg2, …, argn)
```

iterable：一个可迭代对象，如列表或元组。

iterable:An iterable object, such as a list or tuple.

arg1, arg2, …, argn：多个参数，max()会返回这些参数中的最大值。

arg1, arg2, ⋯, argn: Multiple arguments, where max() will return the maximum value among these arguments.

```
numbers=(1, 3, 7, 9, 2, 8)
print(max(numbers))         #Output: 9
```

```
print(max(1, 3, 7, 9, 2, 8))     #Output: 9
```

（6）min()。min()是 Python 中的一个内置函数，它的功能与 max() 函数类似，但是 min() 返回的是可迭代对象中的最小值，或者多个参数中的最小值。语法：

(6)min(). The min() function in Python is a built-in function similar to max(), but it returns the minimum value in an iterable object or among multiple arguments. Syntax:

```
min(iterable)
min(arg1, arg2, …, argn)
```

例如：

For example:

```
numbers=[1, 3, 7, 9, 2, 8]
print(min(numbers))        #Output:1
```

```
print(min(1, 3, 7, 9, 2, 8))   #Output:1
```

（7）sum()。在 Python 中，sum() 是一个内置函数，用于计算可迭代对象（如列表、元组等）中所有元素的和。

(7)sum(). sum() is a built-in function in Python used to calculate the sum of all elements in an iterable object (such as a list, tuple, etc.).

```
numbers=[1, 2, 3, 4, 5]
total=sum(numbers)
print(total)   #Output: 15
```

（8）查找元素。要检查一个元素是否在序列中，可以使用成员运算符 in 或 not in。此外，还可以使用 index() 方法查找元素在列表中的索引（如果元素存在的话）。

(8)Finding elements. To check if an element is in a sequence, one can use the membership operators in or not in. Additionally, one can use the index() method to find the index of an element in a list(if the element exists).

```
my_tuple=(10, 20, 30, 40, 50)
print(30 in my_tuple)         #Output: True
print(60 not in my_tuple)     #Output: True
```

index() 方法是序列类型的一种常用方法，用于查找指定元素在序列中首次出现的位置。如果找不到该元素，则会引发 ValueError 异常。语法：

index() function is a common method for sequence types, used to find the first occurrence of a specified element in the sequence. If the element is not found, it will raise a ValueError exception. Syntax:

```
seq.index(value[, start[, end]])
```

第5章 列表与元组
Chapter 5　List and Tuple

seq：表示原始序列。
value：要查找的元素。
start 和 end（可选参数）：查找的范围，即在一个切片 seq[start：end]中进行查找。

seq: Represent the original sequence.
value: The element to be found.
start and end (optional parameters): The range of search, i. e. searching within a slice seq[start:end].

```
my_tuple=(10, 20, 30, 40, 50)
index=my_tuple.index(20)    #Find the index of element 20
print(index)                #Output: 1
```

（9）遍历序列。可以使用 for 循环来遍历序列中的每个元素。

(9)Iterating through a sequence. One can use a for loop to iterate through each element in the sequence.

```
my_tuple=(10, 20, 30, 40, 50)
for element in my_tuple:
    print(element)
'''
Output:
10
20
30
40
50
'''
```

（10）排序。对于列表，可以使用 sort()方法进行升序或者降序排序，或使用 reverse()方法进行序列的逆序排序。这些方法均会修改列表本身。对于字符串和元组，可以使用内置的 sorted()函数来获取一个排序后的新列表，或使用切片操作来进行逆序排序。但这些操作不会修改原始的字符串或元组。
对于列表：

(10)Sorting. For lists, one can use the sort() method for ascending or descending order sorting, or use the reverse() method for reversing the sequence. These methods will all modify the list itself. For strings and tuples, one can use the built-in sorted() function to get a sorted new list, or use slicing operations for reverse sorting. However, these operations will not modify the original string or tuple.
For lists:

```
my_list=[50, 20, 30, 10, 40]
my_list.sort()              #Ascending sorting
print(my_list)              #Output:[10, 20, 30, 40, 50]
my_list.reverse()           #Descending sorting of a list
print(my_list)              #Output: [50, 40, 30, 20, 10]
```

对于元组或字符串： For tuples or strings:

```
#For tuple operations
my_tuple=(3,2,1,5,4)
sorted_tuple=sorted(my_tuple)        #Ascending sorting of tuple
print(sorted_tuple)                   #Output: [1, 2, 3, 4, 5]
#Because tuples cannot be changed, they will be converted into a list after sorting.
```

（11）连接运算符+：用于连接两个序列。

(11)Concatenation Operator +: Used to concatenate two sequences.

```
tuple1=(1, 2, 3)
tuple2=(4, 5, 6)
print(tuple1+tuple2)   #Output(1, 2, 3, 4, 5, 6)
```

（12）重复运算符 * ：用于重复序列的内容。

(12) Repetition Operator *: Used to repeat the contents of a sequence.

```
my_tuple=(1, 2, 3)
print(my_tuple*3)   #Output(1, 2, 3, 1, 2, 3, 1, 2, 3)
```

（13）比较运算符（==，！=，<，>，<=，>=）：用于比较序列中的元素。序列的比较是逐个元素进行的，按照元素顺序进行比较。

(13)Comparison Operators(==, !=, <, >, <=, >=): Used to compare elements in a sequence. Comparison of sequences is done element by element, following the order of the elements.

```
tuple1=(1, 2, 3)
tuple2=(1,2,3,4)
print(tuple1==tuple2)          #Output: False
print(tuple1<=tuple2)          #Output: True
```

5.4 序列综合应用
5.4 The Application of Sequence

（1）split()。在 Python 中，split()是一种字符串类型的方法，用于将字符串按照指定的分隔符拆分成多个子字符串，并返回一个包含这些子字符串的列表。下面是 split()的语法：

(1)split(). In Python, split() is a method of the string data type used to split a string into multiple substrings based on a specified delimiter, and returns a list containing these substrings. Here is the syntax of split():

```
str.split([sep[, maxsplit]])
```

str 参数：要分割的字符串。

sep 参数：一个可选参数，指定用于分割字符串的分隔符。如果没有指定该参数或设为 None，则方法会分割字符串中的所有空白字符（包括空格、换行符 \n、制表符 \t 等）。

maxsplit 参数：一个可选参数，指定最大分割次数。如果指定了该参数，则分割将不会超过这个指定的次数，结果中将包含剩余未分割的部分。

【例 5-7】split()示例。

str parameter: The string to be split.

sep parameter: An optional parameter that specifies the delimiter to use for splitting the string. If the sep is not specified or set to None, the method will split the string based on all whitespace characters (including spaces, newline \n, tab \t, etc.).

maxsplit parameter: an optional parameter that specifies the maximum number of splits to make. If this parameter is specified, the splitting will not exceed this number, and the result will include the remaining unsplit part.

[Example 5-7] Example of split().

```
#Using the default delimiter(whitespace characters)
text="Hello World   This is a test"
words=text.split()           #Equivalent to text.split(None)
print(words)                 #Output:['Hello', 'World', 'This', 'is', 'a', 'test']

#Using a specified delimiter
text="apple,banana,cherry,dates"
fruits=text.split(",")
print(fruits)                #Output:['apple', 'banana', 'cherry', 'dates']
#Specifying the maximum number of splits
text="apple,banana,cherry,dates"
limited_fruits=text.split(",", 2)
print(limited_fruits)        #Output:['apple', 'banana', 'cherry,dates']

#Using a different delimiter and specifying the maximum number of splits
text="apple|banana,cherry;dates"
mixed_fruits=text.split("|", 1)
print(mixed_fruits)          #Output:['apple', 'banana,cherry;dates']
```

请注意，在调用 split() 方法时，原始字符串 str 不会被修改，而是返回一个新的列表。

如果 maxsplit 被指定，则分割操作将在达到指定次数后停止，并且剩余的部分将作为列表的最后一个元素返回。如果 sep 未被指定或

Please note that when calling the split() method, the original string str is not modified; instead, a new list is returned.

If maxsplit is specified, the splitting operation will stop after the specified number of splits, and the remaining part will be returned as the last element in the list. If sep is not specified or is None, whitespace charac-

为 None，则空白字符将被用作分隔符。如果 sep 是一个空字符串，则每个字符都将被分割成一个单独的列表元素。

（2）join()。在 Python 中，join() 是一种字符串类型的方法，它将一个包含多个字符串的迭代对象（例如列表或元组）连接成一个单独的字符串。在连接过程中，可以使用指定的分隔符将各个元素分隔开。语法如下：

ters will be used as the delimiter. If sep is an empty string, each character will be split into a separate list element.

(2)join(). In Python, join() is a method for strings that is used to concatenate multiple strings contained in an iterable object(such as a list or tuple) into a single string. During the concatenation process, a specified separator can be used to separate each element. The syntax is as follows:

> str.join(iterable)

str 参数：一个字符串，指定了将用作分隔符的字符或字符序列。

iterable 参数：一个可迭代的对象，例如列表或元组，其中包含将要连接的字符串元素。

join() 方法返回一个字符串，它是通过将 iterable 中的元素以指定的分隔符连接起来而创建的。

【例 5-8】下面是一些使用 join() 方法的例子。

str parameter: a string that specifies the character or character sequence to be used as the separator.

iterable parameter: an iterable object, such as a list or tuple, that contains the string elements to be joined.

The join() method returns a string created by concatenating the elements of an iterable with a specified separator.

[Example 5-8] Here are some examples of using the join() method.

```
#Joining a list of strings with commas
str_list=['apple', 'banana', 'cherry']
joined_str=', '.join(str_list)
print(joined_str)               #Output:'apple, banana, cherry'

#Joining a list of characters with an empty string
chars=['a', 'b', 'c']
joined_chars=''.join(chars)
print(joined_chars)             #Output:'abc'

#Joining multiple lines of text with a newline character
lines=['This is line 1', 'This is line 2', 'This is line 3']
joined_lines='\n'. join(lines)
```

5-8

```
print(joined_lines)
'''
Output:
This is line 1
This is line 2
This is line 3
'''
#Attempting to join non-string elements will result in a TypeError
num_list=[1, 2, 3]
#The following line of code will raise an error because the elements in the list are not strings
joined_nums=','.join(num_list)   #TypeError: sequence item 0: expected str instance, int found
```

在使用 join()方法时，确保传递给它的迭代对象中的每个元素都是字符串类型。如果需要连接其他类型的元素，则必须先将它们转换为字符串，再调用 join()方法。

When using the join() method, make sure that each element in the iterable object passed to it is of string type. If one needs to concatenate elements of other types, one must first convert them to strings before calling the join() method. One can use the map function to convert integers to strings.

5.5 本章小结
5.5 Summary of This Chapter

在 Python 中，列表和元组都是非常重要的序列类型，它们用于存储有序的元素。以下是本章内容总结。

列表：

（1）列表是可变的，意味着可以在创建列表后修改、添加或删除其元素。

（2）列表的大小是动态的，可以使用一些函数方法对其进行长度和内容的改变。

（3）列表中的元素可以通过索引访问，每一个元素有两个索引编码：正向索引（从 0 开始到序列长度减 1 结束，方向为从左至右）和逆向索引（从-1 开始到列表长度取负值结束，方向为从右至左）。

（4）列表支持大量的内置方法，

In Python, lists and tuples are both essential sequence types used for storing ordered elements. Here is the summary of this chapter.

List:

(1)Lists are mutable, which means that the elements in a list can be modified, added, or removed after the list is created.

(2)The size of a list is dynamic, and one can use various methods to change its length and content.

(3)The elements in a list can be accessed by index, where each element has two index codes: positive indexing(starting from 0 to the sequence length minus 1, in a left-to-right direction) and negative indexing (starting from −1 to the negative length of the list, in a right-to-left direction).

(4)Lists support a variety of built-in methods, such

如 append()、insert()、remove()、pop()和sort()等。

（5）列表可以包含不同数据类型的元素，包括列表类型（嵌套列表）。

元组：

（1）元组是不可变的，一旦创建，不能修改其内容（但可以修改元组中可变对象的内部状态）。例如，当元组中的元素有列表类型的数据时，可以修改这个列表的内容。

（2）与列表不同，由于元组是不可变的序列类型，所以其长度保持不变（尽管它可以包含可变长度的元素，如列表）。

（3）与列表类似，元组中的元素也可以通过索引访问，索引方式和编码方法同列表。

（4）由于元组的不可变性，它支持的方法较少，主要包括count()、index()和一些运算符操作。元组可以包含不同类型的对象。

列表和元组都是 Python 中用于处理序列的数据类型，本章介绍的序列通用操作在字符串、元组和列表中均适用。

as append(), insert(), remove(), pop(), and sort(), etc.

(5)Lists can contain elements of different data types, including list types(nested lists).

Tuple:

(1)Tuples are immutable, meaning that once created, their contents cannot be changed(though one can modify the internal state of mutable objects within the tuple). For example, if a tuple contains a list, one can modify the content of this list.

(2)Unlike lists, since tuples are immutable sequence types, their length remains constant(although they can contain elements of variable length, such as lists).

(3)Similar to lists, elements in a tuple can also be accessed by index, using the same indexing and encoding methods as lists.

(4)Due to the immutability of tuples, they support fewer methods, mainly including count(), index(), and some operator operations. Tuples can contain objects of different types.

In Python, lists and tuples are both data types used for handling sequences. The common sequence operations introduced in this chapter are applicable to strings, tuples, and lists.

5.6 本章练习
5.6 Exercise for This Chapter

5.1 使用 range()函数生成 0~50、间隔为 5 的等差数列，再使用 join()函数将上述数列的数字输出成如下格式：0，5，10，15，…，50。使用列表推导式实现。

5.2 给定一个日期字符串"2024-03-15"，使用 split()函数将其拆分为年、月、日，并使用 join()

5.1 Generate an arithmetic sequence from 0 to 50 with an interval of 5 using the range() function, then use the join() function to output the numbers in the format: 0, 5, 10, 15,…, 50. Implement this using list comprehension.

5.2 Given a date string "2024-03-15", use the split() function to split it into year, month, and day, and then use the join() function to reformat it into the format

函数将其重组为"15/03/2024"的格式。

5.3 给定一个字符串"apple, 5.2, 10; banana, 2.99, 15; orange, 3.49, 8",使用 split() 函数将其拆分为多个项目(用";"进行拆分),并将每个项目的价格和数量相乘,最后将所有结果使用 join() 函数连接成一个由逗号分隔的字符串。

5.4 给定一个字符串"Hello world from Python",使用 split() 函数将其拆分为单词列表,然后使用 join() 函数将这些单词用"-"连接起来。

5.5 编写程序,删除列表 [1, 2, 2, 3, 4, 4, 5, 5, 3] 中的重复值。

5.6 给定一个列表 ['a', 'b', 'c', 'd', 'e'],使用 for 循环实现将列表元素循环左移一个位置,结果是 ['b', 'c', 'd', 'e', 'a']。

5.7 数字 202 从左往右和从右往左读都是一样的,这种数被称为回文数。使用 for 循环及切片设计程序,实现找出 1 000 以内的所有回文数。

5.8 找出整数列表 [4, 2, 9, 7, 5, 9, 3, 9] 中最大元素的下标,如果最大元素的个数超过 1,则打印输出所有下标。

5.9 给定一个包含学生分数的元组列表,代码如下所示,每个学生有多门科目的分数。找出拥有最高平均分的学生,并输出该学生的所有科目分数和平均分。如果有多个学生拥有相同的最高平均分,则输出所有这些学生的信息。

"15/03/2024".

5.3 Given a string "apple,5.2,10;banana,2.99,15;orange,3.49,8", use the split() function to split it into multiple items(split by ";"), multiply the price and quantity of each item, and then join all the results into a commaseparated string using the join() function.

5.4 Given the string "Hello world from Python", use the split() function to split it into a list of words and then use the join() function to connect these words using "-".

5.5 Write a program to remove duplicate values from the list [1, 2, 2, 3, 4, 4, 5, 5, 3].

5.6 Given the list ['a', 'b', 'c', 'd', 'e'], use a for loop to implement shifting the elements of the list one position to the left, resulting in ['b', 'c', 'd', 'e', 'a'].

5.7 Numbers that read the same from left to right and right to left, such as 202, are called palindromic numbers. Design a program using a for loop and slicing to find all palindromic numbers within 1,000.

5.8 Find the index of the maximum element in the integer list [4, 2, 9, 7, 5, 9, 3, 9]. If the maximum element appears more than once, then print out all the indices.

5.9 Given a list of tuples containing student scores as shown in the code below, where each student has scores for multiple subjects. Find the student(s) with the highest average score and output all the subject scores and the average score for that student. If there are multiple students with the same highest average score, output the information for all those students.

```
'''
The list of tuples containing student scores is formatted as(student name, subject 1 score, subject 2 score, …)
'''
students_scores=[
    ("Alice", 85, 90, 78),
    ("Bob", 92, 88, 90),
    ("Charlie", 76, 82, 80),
    ("David", 92, 90, 94),
    ("Eve", 88, 85, 87)
]
```

本章练习题参考答案

第 6 章 字典与集合
Chapter 6 Dictionaries and Sets

6.1 字典的定义
6.1 Definition of Dictionaries

在 Python 中，字典是一种可变的数据结构，它存储键值对。字典中的每个键都是唯一的，并与一个值相关联。可以使用键来访问、修改或删除字典中的值，字典结构示意图如图 6-1 所示。

In Python, a dictionary is a mutable data structure that stores key-value pairs. Each key in the dictionary is unique and associated with a value. Keys can be used to access, modify, or delete values in the dictionary, the dictionary structure diagram shown in Figure 6-1.

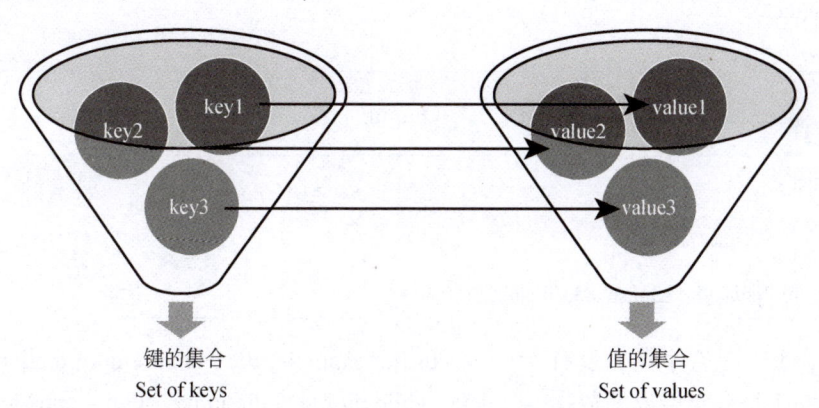

图 6-1 字典结构示意图

Figure 6-1 Dictionary Structure Diagram

字典具有如下特性。

（1）字典的元素是键值对，字典中的键不允许重复，必须有唯一值，而且键不可变。

（2）字典不支持索引和切片操

Dictionaries have the following characteristics.

(1) Dictionary elements are key-value pairs, where keys must be unique and immutable in the dictionary.

(2) Dictionaries do not support indexing and slicing

165

作，但可以通过键查询值。

（3）字典是无序的对象集合。

（4）字典是可变的，并且值可以是任何 Python 对象，也可以是字典类型。

operations, but values can be queried by keys.

(3)Dictionaries are unordered collections of objects.

(4)Dictionaries are mutable, and values can be any Python object, including dictionary types.

6.2 字典的创建
6.2 Creating Dictionaries

6.2.1 使用大括号创建字典
6.2.1 Creating Dictionaries Using Curly Braces

创建字典可将键值对放在大括号"{}"中，多个键值对使用逗号","分隔。语法格式如下：

字典名称 = {键1:值1, 键2:值2, …, 键n:值n}

To create a dictionary, the key-value pairs are enclosed in curly braces "{}", with multiple key-value pairs separated by commas ",". The syntax format is as follows:

Dictionary_name = {Key1:Value1, Key2:Value2, …, Key n:Value n}

```
dic1 = {}
dic2 = {'python':99, 'java':98, 'english':88, 'math':95}
print(type(dic1))
print(dic1)
print(type(dic2))
print(dic2)
```

输出：

Output:

```
<class 'dict'>
{}
<class 'dict'>
{'python':99, 'java':98, 'english':88, 'math':95}
```

示例中，dic1 是空字典，没有任何数据，dic2 是有 4 条成绩数据的字典。字典的键为科目，字典的值为科目对应的成绩。

In the example, dic1 is an empty dictionary with no data, while dic2 is a dictionary with 4 entries of grade data. The keys of the dictionary represent subjects, and the values represent the corresponding grades for each subject.

6.2.2 使用 dict() 函数创建字典
6.2.2 Creating Dictionaries Using the dict() Function

在 Python 中，dict() 是一个内置函数，用于创建一个新的字典对

In Python, dict() is a built-in function used to create a new dictionary object. Similar to list() function and

象。dict()函数的使用与 list()函数、tuple()函数类似，直接使用它将创建一个空字典。

tuple() function, using dict() directly will create an empty dictionary.

```
empty_dict=dict()
print(type(empty_dict))
print(len(empty_dict))
print(empty_dict)
```

输出：　　　　　　　　　　　　Output:

```
<class 'dict'>
0
{}
```

【例 6-1】dict()函数支持使用包含两个元素的元组（或其他可迭代对象）序列来创建字典，其中，每个元组的第一个元素是键，第二个元素是值。它还支持使用关键字参数创建字典、从另一个字典创建新的字典。

[Example 6-1] The dict() function supports creating a dictionary by using a sequence containing tuples with two elements(or other iterable objects), where the first element of each tuple is the key and the second element is the value. It also supports creating dictionaries using keyword arguments and creating a new dictionary from another dictionary.

```
#Create dictionaries by using tuples with two elements
dict_from_sequence=dict([(1, 'one'),(2, 'two'),(3, 'three')])
print(dict_from_sequence)
dict_from_keywords=dict(name='Li Ming', age=20, city='Shenyang')
#Create dictionaries by using keyword arguments
print(dict_from_keywords)
another_dict={'a': 1, 'b': 2}
new_dict_from_mapping=dict(another_dict)
print(new_dict_from_mapping)
```

6-1

输出：　　　　　　　　　　　　Output:

```
{1: 'one', 2: 'two', 3: 'three'}
{'name': 'Li Ming', 'age': 20, 'city': 'Shenyang'}
{'a': 1, 'b': 2}
```

还可以使用 fromkeys()方法创建具有相同默认值的字典。

One can also use the fromkeys() method to create a dictionary with the same default value.

```
dic=dict.fromkeys('Python',0)
print(dic)
```

输出：　　　　　　　　　　　　　　Output:

{'P': 0, 'y': 0, 't': 0, 'h': 0, 'o': 0, 'n': 0}

6.2.3 使用 map()函数创建字典
6.2.3 Creating Dictionaries Using the map() Function

在 Python 中，map()函数是一个内置函数，用于将指定的函数应用于可迭代对象（如列表、元组等）中的每一个元素，并返回一个新的迭代器，该迭代器产生原序列中每个元素经过函数处理后的结果。当然也可以使用 map()函数代替推导式来创建字典。

In Python, the map() function is a built-in function used to apply a specified function to each element of an iterable object(such as a list, tuple, etc.) and return a new iterator that yields the results of the function applied to each element of the original sequence. map() function can also be used as an alternative to comprehensions to create dictionaries.

```
dic=map(lambda num:[num,chr(num)],range(97,123))
print(dic)
print(dict(dic))
```

输出：　　　　　　　　　　　　　　Output:

```
<map object at 0x0000024761947100>
{97: 'a', 98: 'b', 99: 'c', 100: 'd', 101: 'e', 102: 'f', 103: 'g', 104: 'h', 105: 'i', 106: 'j', 107: 'k', 108: 'l',
109: 'm', 110: 'n', 111: 'o', 112: 'p', 113: 'q', 114: 'r', 115: 's', 116: 't', 117: 'u', 118: 'v', 119: 'w',
120: 'x', 121: 'y', 122: 'z'}
```

示例中，使用匿名函数将 range 对象表示的等差数列 97 到 122 中的每一个数 num 转换成［num，chr（num）］的形式，并返回可迭代的 map 对象，再使用 dict（）函数将 map 对象转换成字典。

In the example, using an anonymous lambda function, each number num in the arithmetic sequence represented by the range object from 97 to 122 is transformed into the form [num, chr(num)]. This transformation results in an iterable map object. Then, the map object is converted into a dictionary using the dict() function.

6.2.4 使用推导式创建字典
6.2.4 Creating Dictionaries Using Comprehensions

在 Python 中，字典推导式是一种创建字典的简洁方法。它使用类似于列表推导式的语法，但返回的是一个字典，而不是列表。字典

In Python, dictionary comprehension is a concise way to create dictionaries. It uses a syntax similar to list comprehension but returns a dictionary instead of a list. The basic syntax of dictionary comprehension is as fol-

Chapter 6　Dictionaries and Sets

推导式的基本语法如下：

lows:

{key_expression: value_expression for item in iterable [if condition]}

其中：

（1）key_expression 用于生成字典的键；

（2）value_expression 用于生成字典的值；

（3）item 是从 iterable 中取出的元素；

（4）condition 是一个可选的表达式，用于过滤元素。

【例6-2】使用推导式构造字典，输出 Unicode 码从 9800 开始到 9811 对应的星座符号。

Where:

(1) key_expression is used to generate the keys of the dictionary;

(2) value_expression is used to generate the values of the dictionary;

(3) item is an element extracted from the iterable;

(4) condition is an optional expression used for filtering elements.

[Example 6-2] Use the dictionary comprehension to construct a dictionary that outputs the zodiac symbols corresponding to the Unicode characters from 9800 to 9811.

```
#Use the dictionary comprehension to construct a dictionary
dic = {num:chr(num) for num in range(9800,9812)}
print(dic)
```

输出：

Output:

{9800: '♈', 9801: '♉', 9802: '♊', 9803: '♋', 9804: '♌', 9805: '♍', 9806: '♎', 9807: '♏', 9808: '♐', 9809: '♑', 9810: '♒', 9811: '♓'}

另一个示例：

Another example:

```
keys = ['name', 'age', 'city']
values = ['Li Ming', 20, 'Shenyang']
dic = {k: v for k, v in zip(keys, values)}
print(dic)
```

输出：

Output:

{'name': 'Li Ming', 'age': 20, 'city': 'Shenyang'}

6.2.5　使用 collections 模块中的类创建字典
6.2.5　Creating Dictionaries Using Classes in the Collections Module

在 collections 模块中有 3 个类可以用来创建字典，包括 defaultdict、

In the collections module, there are three classes that can be used to create dictionaries, including defaultdict,

OrderedDict 和 Counter。

（1）defaultdict 是 Python 标准库 collections 中的一个类，它提供了一种字典子类，为字典的 get() 方法提供了默认值。

当访问字典中不存在的键时，defaultdict 会自动为该键提供一个默认值，而不是引发 KeyError 异常。要使用 defaultdict，需要先导入 collections 模块，并提供一个工厂函数作为 defaultdict 的第一个参数，这个工厂函数描述了当字典访问一个不存在的键时应该返回什么默认值。

OrderedDict and Counter.

(1) defaultdict is a class in the Python standard library collections. It provides a subclass of dictionary that offers a default value for the get() method of the dictionary.

When accessing a key that does not exist in the dictionary, defaultdict automatically provides a default value for that key instead of raising a KeyError exception. To use defaultdict, one needs to import the collections module and provide a factory function as the first parameter of defaultdict. This factory function describes what default value should be returned when the dictionary accesses a non-exist key.

```
from collections import defaultdict
counter = defaultdict(lambda : 0)
counter['apple'] += 1
counter['banana'] += 1
counter['apple'] += 1
print(counter)
```

输出：

Output:

```
defaultdict(<function<lambda> at 0x0000022D4427BF60>, {'apple': 2, 'banana': 1})
```

（2）OrderedDict 类似于正常的字典，它记住了元素插入时的顺序，当在有序的字典上迭代时，按第一次添加元素的顺序返回元素，这样 OrderedDict 就是一个有序的字典。

(2) OrderedDict is similar to a regular dictionary, it remembers the order in which elements were inserted. When iterating over an OrderedDict, the elements are returned in the order they were first added. This makes OrderedDict an ordered dictionary.

```
from collections import OrderedDict
dic1 = OrderedDict()
dic1['k1'] = 'v1'
dic1['k3'] = 'v3'
dic1['k2'] = 'v2'
print(dic1)
```

输出：

Output:

```
OrderedDict({'k1': 'v1', 'k3': 'v3', 'k2': 'v2'})
```

（3）Counter（计数器）以字典的形式返回序列中各个字符出现的次数，值为 key，次数为 value。Counter 是对字典类型的补充，提供了一个快速而灵活的方式来统计元素出现的次数。

(3) Counter returns a dictionary-like object that counts the occurrences of each character in a sequence, where the characters are the keys and the counts are the values. Counter serves as a complement to the dictionary type, offering a fast and flexible method to count the occurrence of elements.

```
from collections import Counter
counter1 = Counter('hello world')
counter2 = Counter('hello world hello world hello python'.split())
counter3 = Counter(['apple', 'banana', 'apple', 'orange', 'banana', 'banana'])
print(counter1)
print(counter2)
print(counter3)
```

输出： | Output:

```
Counter({'l': 3, 'o': 2, 'h': 1, 'e': 1, ' ': 1, 'w': 1, 'r': 1, 'd': 1})
Counter({'hello': 3, 'world': 2, 'python': 1})
Counter({'banana': 3, 'apple': 2, 'orange': 1})
```

6.3 字典的访问
6.3 Accessing Dictionaries

6.3.1 使用键访问
6.3.1 Accessing by Keys

字典的元素是按照键值对存储的，如果想要得到元素的值，可以通过键来访问字典中的值。

The elements of a dictionary are stored as key-value pairs, if one wants to access the value of an element, one can do so by using the key to access the corresponding value in the dictionary.

```
dic1 = {'name': 'Li Ming', 'age': 20, 'city': 'Shenyang'}
print(dic1['name'])
print(dic1['age'])
print(dic1['city'])
```

输出： | Output:

```
Li Ming
20
Shenyang
```

如果尝试访问一个不存在的键,则 Python 会抛出一个 KeyError 异常。为了避免出现这种情况,可以使用 get() 方法。如果键不存在,则 get() 方法会返回 None(或者可以提供一个默认值作为 get() 方法的第二个参数)。

If one tries to access a non-exist key, Python will raise a KeyError exception. To avoid this situation, one can use the get() method. If the key does not exist, the get() method will return None(or one can provide a default value as the second parameter of the get() method).

```
dic1 = {'name': 'Li Ming',   'age': 20,   'city': 'Shenyang'   }
print(dic1.get('name'))
print(dic1.get('score'))
print(dic1.get('phone','No phone number'))
```

输出:

Output:

```
Li Ming
None
No phone number
```

对于字典中的元素是字典的情况,按照读取字典元素的基本格式使用即可。

For elements in a dictionary that are themselves dictionaries, one can access them using the basic dictionary element access format as needed.

```
dic1 = {
    'Li Ming':{'Python':99,'English':98,'Math':95},
    'Zhang Hong':{'Python':98,'English':89,'Math':78},
    'Jiang Nan':{'Python':85,'English':88,'Math':96}
    }
print(dic1['Li Ming']['Python'],dic1['Li Ming']['English'],dic1['Li Ming']['Math'])
```

输出:

Output:

```
99 98 95
```

6.3.2 访问全部的键和值
6.3.2 Retrieve All Keys and Values

Python 为字典访问提供了 3 个常用方法 keys()、values() 和 items(),可以用来分别获取字典的键、值和键值对,也可以结合 for 循环遍历字典的键、值和键值对。

Python provides three commonly used methods for dictionary access: keys(), values(), and items(). These can be used to respectively retrieve the keys, values, and key-value pairs of a dictionary. One can also combine these methods with a for loop to iterate over the keys, values, and key-value pairs of a dictionary.

【例 6-3】使用 keys()、values() 和 items() 方法遍历访问字典中的键、值和键值对。

[Example 6-3] Use the keys(), values() and items() methods to iterate over a dictionary's keys, values, and key-value pairs.

```
dic1 = {'name': 'Li Ming',   'age': 20,   'city': 'Shenyang'   }
print(dic1.keys())
print(dic1.values())
print(dic1.items())
#Iterate over dic1's keys
for i in dic1.keys():
    print(i)
#Iterate over dic1's values
for i in dic1.values():
    print(i)
#Iterate over dic1's key-value pairs
for k,v in dic1.items():
    print('key:{},value:{}'.format(k,v))
```

输出：

Output:

```
dict_keys(['name', 'age', 'city'])
dict_values(['Li Ming', 20, 'Shenyang'])
dict_items([('name', 'Li Ming'),('age', 20),('city', 'Shenyang')])
name
age
city
Li Ming
20
Shenyang
key:name,value:Li Ming
key:age,value:20
key:city,value:Shenyang
```

6.4 字典元素的添加、修改和删除
6.4 Adding, Modifying, and Deleting Dictionary Elements

添加和修改字典元素的操作相同，基本格式为：dict[key] = new_value。如果 key 在字典 dict 中存在，则用新值 new_value 替换键对应的旧值，相当于修改操作，否则在字典 dict 中添加新元素 key-

The operation for adding and modifying dictionary elements is the same, with the basic format: dict[key] = new_value. If the key exists in the dictionary dict, the new_value will replace the old value corresponding to the key, equivalent to a modification operation. Otherwise, a new element key-new_value pair will be

173

new_value 对。 | added to the dictionary dict.

```
score={'Python':99,'English':98,'Math':95}
print(score)
score['Java']=87
print(score)
score['Python']=100
print(score)
```

输出： | Output:

```
{'Python': 99, 'English': 98, 'Math': 95}
{'Python': 99, 'English': 98, 'Math': 95, 'Java': 87}
{'Python': 100, 'English': 98, 'Math': 95, 'Java': 87}
```

删除字典中的某个元素，可以使用 del 关键字，具体的格式为：del dict[key]，表示删除字典 dict 中键为 key 的键值对，如果键 key 存在，则删除键 key 对应的值，否则返回 KeyError 异常。

另外，del 关键字不仅可以用来删除字典中的元素，也可以用来删除整个字典，具体格式为：del 字典名。

To delete a specific element from a dictionary, one can use the del keyword with the format: del dict[key], which signifies deleting the key-value pair with the key in the dictionary dict. If the key exists, the corresponding value will be deleted; otherwise, a KeyError exception will be raised.

Additionally, the "del" keyword can not only be used to delete elements from a dictionary but also to delete the entire dictionary with the format "del dictionary_name".

```
score={'Python':99,'English':98,'Math':95}
print(score)
del score['Math']
print(score)
del score
print(score)
```

输出： | Output:

```
{'Python': 99, 'English': 98, 'Math': 95}
{'Python': 99, 'English': 98}
Traceback(most recent call last):
    File "C:\test.py", line 6, in<module>
        print(score)
NameError: name 'score' is not defined
```

通过输出可以看到，在删除字典对象 score 后执行 print(score) 语句将产生异常，提示 score 对象没

By observing the output, it can be seen that after deleting the dictionary object "score", executing the print(score) statement will result in an exception indicating that the

第6章 字典与集合
Chapter 6 Dictionaries and Sets

有被定义。

可以通过 pop()和 popitem()方法删除字典中的元素。

pop()方法用于删除字典中指定的键值对,并返回该键对应的值。如果指定的键不存在于字典中,则 pop()方法会抛出一个 KeyError 异常。为了避免出现这种情况,可以为 pop()方法提供一个默认值作为第二个参数,这样,当键不存在时,它会返回这个默认值而不是抛出异常。popitem()方法删除的是最后添加到字典的键值对。

【例6-4】使用 pop()方法删除字典中指定的键值对,并返回该键对应的值。

score object has not been defined.

One can use the pop() and popitem() methods to delete elements from a dictionary.

The pop() method is used to remove the specified key-value pair from the dictionary and returns the value corresponding to that key. If the specified key does not exist in the dictionary, the pop() method will raise a KeyError exception. To avoid this situation, one can provide a default value as the second parameter to the pop() method, so that when the key does not exist, it will return this default value instead of raising an exception. The popitem() method removes the key-value pair that was last added to the dictionary.

[Example 6-4] Use the pop() method to remove the specified key-value pair from a dictionary and returns the value associated with the key.

```
score = {'Python':99,'English':98,'Math':95}
print(score)
print(score.pop('Python'))        #Return 99
print(score)
print(score.pop('Java',0))        #Return the default value 0 as the key 'Java' does not exist
print(score.pop('DataBase'))      #Raise an exception as the key 'DataBase' does not exist
```

6-4

输出:

Output:

```
{'Python': 99, 'English': 98, 'Math': 95}
99
{'English': 98, 'Math': 95}
0
Traceback(most recent call last):
   File "C:\test. py", line 6, in<module>
     print(score. pop('DataBase'))
KeyError: 'DataBase'
```

对于最后一行代码 print(score.pop('DataBase')),由于字典中不存在键"DataBase",所以产生异常。

For the last line of code, print(score.pop('DataBase')), will raise an exception because the key "DataBase" does not exist in the dictionary.

```
score = {'Python':99,'English':98,'Math':95}
print(score)
print(score.popitem())
print(score)
```

175

输出： | Output:

```
{'Python': 99, 'English': 98, 'Math': 95}
('Math', 95)
{'Python': 99, 'English': 98}
```

6.5 字典的更新与排序
6.5 Updating and Sorting the Dictionary

如果有两个字典，需要将一个字典的键值对更新到另一个字典中，则可以使用 update() 方法完成。如果两个字典有相同的键，那么新字典的值会覆盖旧字典的值。update() 方法会直接修改原来的字典，而不是返回一个新的字典。如果不想修改原字典，则可以先复制一份再更新。

If one has two dictionaries and needs to update the key-value pairs from one dictionary to another, one can use the update() method to achieve this. If the two dictionaries have the same key, the value in the new dictionary will overwrite the value in the old one. The update() method directly modifies the original dictionary instead of returning a new one. If one does not want to modify the original dictionary, one can first make a copy and then update it.

```
dict1 = {'Python':99,'Java':98}
dict2 = {'Python':95,'English':98}
dict1.update(dict2)
print(dict1)
```

输出： | Output:

```
{'Python': 95, 'Java': 98, 'English': 98}
```

示例中，字典 dict1 和 dict2 中均有键"Python"，当使用 update() 方法更新时，字典 dict1 中的键"Python"对应的值被更新为 95，dict2 中键值对"English：98"被添加到字典 dict1 中。

In the example, dictionary dict1 and dict2 both has the key "Python", when using the update() method to update, the value corresponding to the key "Python" in dictionary dict1 will be updated to 95, and the key-value pair "English: 98" from dict2 will be added to dictionary dict1.

由于字典是无序的键值对集合，所以直接对字典进行排序是没有意义的。但可以对字典的键或值进行排序，或者根据值对键进行排序，使用 sorted() 方法并将字典对象作为其参数来完成排序。

As dictionaries are unordered collections of key-value pairs, directly sorting a dictionary does not make sense. However, one can sort the keys or values of a dictionary, or sort the keys based on their values. This can be achieved using the sorted() method by passing the dictionary object as the argument.

【例 6-5】使用 sorted() 方法对

[Example 6-5] Use the sorted() method to sort the

字典的键、值进行排序，再根据值对键进行排序。

keys and values of a dictionary, then sort the keys based on the values.

```
dict1 = {'Python': 95, 'Java': 96, 'English': 92}
#Sort the keys of dict1
sorted_keys = sorted(dict1.keys())
print(sorted_keys)
#Sort the values of dict1
sorted_values = sorted(dict1.values())
print(sorted_values)
#Sort the dict1 based on values in ascending order
sorted_items = sorted(dict1.items(), key=lambda item: item[1])
sorted_dict = dict(sorted_items)
print(sorted_dict)
```

输出： | Output:

```
['English', 'Java', 'Python']
[92, 95, 96]
{'English': 92, 'Python': 95, 'Java': 96}
```

示例中，sorted(dict1.keys())对字典的键进行排序，sorted(dict1.values())对字典的值进行排序，返回的数据类型均是列表类型。sorted(dict1.items(), key=lambda item: item[1])对字典数据进行排序，返回的数据类型也是列表类型，再使用dict()函数将结果转换成字典，这里是按照字典的值从小到大排序，使用lambda函数，item[1]即指定按照值排序，如果想按照字典的键排序，则需要写成item[0]。

In the example, sorted(dict1.keys()) sorts the keys of the dictionary, sorted(dict1.values()) sorts the values of the dictionary, and the return type for both is a list. Using sorted(dict1.items(), key=lambda item: item[1]) sorts the dictionary data and also returns a list. Then, using the dict function converts the result back into a dictionary. Here, it sorts the dictionary based on values in ascending order by using a lambda function where item[1] specifies sorting by values. If one wants to sort by keys, one would write it as item[0].

6.6 字典拓展示例
6.6 Dictionary Expansion Examples

【例6-6】统计字符串中每个字符出现的次数。

[Example 6-6] Count the occurrence of each character in a string.

```
def count_chars(s):
    char_count = {}
```

```python
    for c in s:
        if c in char_count:
            char_count[c]+=1
#Check if the character c is already in the char_count dictionary, if it is, then increment its corresponding
#value by 1
        else:
            char_count[c]=1
#If c is not in the dictionary, then add the character to the dictionary and set its value to 1.
    return char_count

text="Hello Python!"
print(count_chars(text))
```

6-6

输出： | Output:

```
{'H': 1, 'e': 1, 'l': 2, 'o': 2, ' ': 1, 'P': 1, 'y': 1, 't': 1, 'h': 1, 'n': 1, '!': 1}
```

【例6-7】对学生的期末考试成绩进行统计，将考试成绩按照优（A）、良（B）、中（C）、及格（D）和不及格（E）进行归类，并统计各分数段的人数。

[Example 6-7] Statistical analysis of students' final exam results. Classify the exam scores as excellent(A), good(B), medium(C), passing(D), and failing(E), and tally the number of students in each score range.

```python
scores=[89,70,49,87,92,84,73,71,78,81,90,38,77,72,81,79,80,82,75,90,54,80,70,68,61]
#dic_groups contains empty lists for different groups(A, B, C, D, E)
dic_groups={c : [] for c in"A B C D E".split()}
for score in scores:
    if score>=90:
        dic_groups['A'].append(score)
    elif score>=80:
        dic_groups['B'].append(score)
    elif score>=70:
        dic_groups['C'].append(score)
    elif score>=60:
        dic_groups['D'].append(score)
    else:
        dic_groups['E'].append(score)
#Store scores in corresponding lists
print(dic_groups)
dic_groups_num={key : len(value) for key,value in dic_groups.items()}
print(dic_groups_num)
#The dic_groups_num contains the number of scores in each group
```

6-7

输出： | Output:

{'A': [92, 90, 90], 'B': [89, 87, 84, 81, 81, 80, 82, 80], 'C': [70, 73, 71, 78, 77, 72, 79, 75, 70], 'D': [68, 61], 'E': [49, 38, 54]}
{'A': 3, 'B': 8, 'C': 9, 'D': 2, 'E': 3}

6.7　集合的定义
6.7　Definition of a Set

Python 中的集合是一个无序且不包含重复元素的集合，是一种内置常用的可迭代数据类型。集合主要用于成员检测以及消除重复元素。集合的基本用法包括添加元素、移除元素、判断元素是否存在于集合中，以及计算集合的交集、并集、差集等。

集合中的元素要求是整数、实数、复数、字符串、元组等不可变数据类型或可进行哈希的数据。列表、字典、集合等可变类型数据不能进行哈希，因此不能作为集合的元素。集合本身是可变的，创建后可以添加、删除元素，元素存储顺序与添加顺序并不一致，所以集合是无序的结构。在集合中没有索引的概念，不支持使用下标直接访问指定位置的元素，也不支持切片、随机选取等操作。

In Python, a set is an unordered collection of unique elements that is commonly used as a built-in iterable data type. Sets are primarily used for membership testing and eliminating duplicate elements. Basic operations with sets include adding elements, removing elements, checking for the existence of elements in the set, computing the intersection, union, and set difference.

Elements in a set must be of immutable data types or hashable data such as integers, floats, complex numbers, strings, tuples, etc. Mutable data types like lists, dictionaries, sets, etc., are not hashable and thus cannot be elements in a set. Sets themselves are mutable, allowing one to add and remove elements after creation. The order of elements in a set is not consistent with the order in which elements are added, making sets unordered structures. Sets do not have the concept of indexing, so one cannot access specific elements by index, slicing, or perform random selection operations.

6.8　集合的创建
6.8　Creating a Set

6.8.1　使用大括号和 set() 函数创建集合
6.8.1　Use Braces and the set() Function to Create Sets

可以使用大括号{}或 set()函数来创建集合，需要注意的是，创建一个空集合必须使用 set()而不是

Sets can be created using curly braces {} or the set() function. It is important to note that creating an empty set requires using set() instead of {}, as {} is used to create

179

{}，因为{}用于创建空字典。如果数据中存在重复元素，则在转换时只保留一个。如果元素中有可变类型的对象，则会抛出 TypeError 异常。

an empty dictionary. If there are duplicate elements in the data, only one will be retained during conversion. If the elements contain mutable objects, a TypeError exception will be raised.

```
set1 = {1, 2, 3, 4}
set2 = set([1, 1, 2, 2, 3, 3, 4, 4])
set3 = set('Hello Python!')
set4 = set()
print(set1)
print(set2)
print(set3)
print(set4)
```

输出：

Output:

```
{1, 2, 3, 4}
{1, 2, 3, 4}
{'y', 'l', ' ', 'o', 'P', 'n', 't', 'e', 'H', '!', 'h'}
set()
```

当使用{}创建集合且元素是可变类型时，会抛出"TypeError: unhashable type"异常。

When creating a set using {}, if the elements are of mutable types, a "TypeError: unhashable type" exception will be raised.

```
set1 = set([1,2,3,4])
print(set1)
set2 = {[1,2,3,4]}
print(set2)
```

输出：

Output:

```
{1, 2, 3, 4}
Traceback(most recent call last):
    File "C:\test.py", line 3, in<module>
        set2 = {[1, 2, 3, 4]}
TypeError: unhashable type: 'list'
```

从运行结果可以看出，set()函数可以将可变数据类型转为集合，而{}是创建集合，其中的元素不能出现可变对象。

From the execution result, it can be seen that the set() function can convert mutable data types into sets, while {} is used to create sets where the elements cannot be mutable objects.

6.8.2 使用推导式创建集合
6.8.2 Use Set Comprehension to Create Sets

在 Python 中，集合推导式是一种根据现有可迭代对象（如列表、元组、字符串或另一个集合）创建新集合的简洁方式。

集合推导式的基本语法与列表推导式类似，但结果是一个集合，因此会自动去除重复元素。集合推导式的基本语法如下：

Set comprehension in Python is a concise way to create a new set from an existing iterable object(such as a list, tuple, string, or another set).

The basic syntax of set comprehension is similar to list comprehension, but the result is a set, so duplicate elements are automatically removed. The basic syntax of set comprehension is as follows:

```
{expression for item in iterable [if condition]}
```

其中：

（1）expression 是对每个 item 进行操作的表达式；

（2）item 是从 iterable 中取出的元素；

（3）iterable 是任意可迭代对象；

（4）condition 是一个可选的表达式，用于过滤 item。

【例 6-8】使用集合推导式创建集合。

Where:

(1)expression is the expression that operates on each item;

(2)item is the element taken from the iterable;

(3)iterable is any iterable object;

(4)condition is an optional expression used to filter items.

[Example 6-8] Use set comprehension to create sets.

```
numbers=[1, 2, 2, 3, 4, 4, 5, 6, 7, 8, 9, 10, 10]
set1 = {num for num in numbers}                    #The set of unique elements in the list 'numbers'
set2 = {num for num in numbers if num % 2==0}      #The set of even elements in the list 'numbers'
set3 = {num**2 for num in numbers}                 #The set of squares of each element in the list 'numbers'
print(set1)
print(set2)
print(set3)
word = "hello python"
set4 = {letter for letter in word}                 #The set of unique letters in the string 'hello python'
print(set4)
```

输出： | Output:

```
{1, 2, 3, 4, 5, 6, 7, 8, 9, 10}
{2, 4, 6, 8, 10}
{64, 1, 4, 36, 100, 9, 16, 49, 81, 25}
{'t', 'n', 'p', 'e', ' ', 'y', 'o', 'l', 'h'}
```

6-8

6.9 集合元素的添加和删除
6.9 Adding and Removing Elements in a Set

可以使用 add() 方法向集合中添加单个元素，被添加的元素只能是字符串、数字和布尔类型的 True 或 False 等，不能是列表、元组等可迭代对象。如果被添加的元素不在集合中，则向集合中添加该元素，否则不进行任何操作。另外，也可以使用 update() 方法向集合中添加多个元素。

One can use the add() method to add a single element to the set. The element being added can only be a string, number, or a Boolean value like True or False, and cannot be a list, tuple, or other iterable objects. If the element being added is not in the set, it will be added; otherwise, no operation will be performed. Alternatively, one can use the update() method to add multiple elements to the set.

```python
my_set=set()
my_set.add(1)
my_set.add(1)
my_set.add(2)
my_set.add(2)
my_set.add('abcd')
my_set.add(False)
print(my_set)
my_set.update([3, 4])
print(my_set)
other_set={5, 6}
my_set.update(other_set)
print(my_set)
```

输出：

Output:

```
{False, 1, 2, 'abcd'}
{False, 1, 2, 3, 4, 'abcd'}
{False, 1, 2, 3, 4, 5, 6, 'abcd'}
```

可以使用 remove() 方法删除集合中的单个元素，或者使用 discard() 方法尝试删除元素（如果元素不存在，也不会引发错误）。还可以使用 pop() 方法随机删除并返回集合中的一个元素，或者使用 clear() 方法清空整个集合。

One can use the remove() method to delete a single element from the set, or use the discard() method to attempt to remove an element (without raising an error if the element does not exist). One can also use the pop() method to randomly remove and return one element from the set, or use the clear() method to empty the entire set.

```
my_set={'p', 'y', 1, 2, 3, 4, 5, 't', 'h', 'o', 'n', 'hello'}
my_set.remove(5)
print(my_set)
my_set. discard(10)
print(my_set)
removed_item=my_set. pop()
print(removed_item)
print(my_set)
my_set. clear()
print(my_set)
```

输出： | Output:

```
{'n', 1, 2, 'h', 3, 4, 'y', 't', 'o', 'p', 'hello'}
{'n', 1, 2, 'h', 3, 4, 'y', 't', 'o', 'p', 'hello'}
n
{1, 2, 'h', 3, 4, 'y', 't', 'o', 'p', 'hello'}
set()
```

6.10 集合的运算
6.10 Set Operations

Python 中的集合与数学中的集合概念是一致的，因此两个集合可以做数学意义上的交集、并集、差集和对称差等运算。这些操作在处理集合数据时非常有用，可以帮助快速合并、比较或过滤集合中的元素。集合的运算如表 6-1 所示。

Sets in Python align with the concept of sets in mathematics, allowing operations such as intersection, union, set difference, and symmetric difference, which have mathematical meanings. These operations are very useful for manipulating set data, enabling quick merging, comparisons, or filtering of elements in sets. Set Operations are illustrated in Table 6-1.

表 6-1 集合的运算
Table 6-1 Set Operations

Method	Function Description
A&B or A. intersection(B)	交。返回一个新集合，其中包含同时存在于集合 A 和集合 B 中的元素。 Intersection. Return a new set that contains elements that are present in both set A and set B
A|B or A. union(B)	并。返回一个新集合，其中包含集合 A 和集合 B 中的所有元素。 Union. Return a new set that contains all elements from both set A and set B
A-B or A. difference(B)	差。返回一个新集合，其中包含集合 A 中不在集合 B 中的元素。 Difference. Return a new set that contains elements from set A that are not present in set B

续表

Method	Function Description
A^B or A.symmetric_difference(B)	对称差。返回一个新集合，其中包含集合 A 和集合 B 中的元素，但不包括同时存在于两个集合中的元素。 Symmetric Difference. Return a new set that contains elements from set A and set B, but does not include elements that are present in both sets at the same time
A<=B or A.issubset(B)	子集测试。如果集合 A 与集合 B 相同或集合 A 是集合 B 的子集，则返回真(True)；否则，返回假(False)。 Subset Test. If A is the same as B or A is a subset of B, return True; otherwise, return False.
A>=B or A.issuperset(B)	超集测试。如果集合 A 与集合 B 相同或集合 A 是集合 B 的超集，则返回真(True)；否则，返回假(False)。 Superset Test. If A is the same as B or A is a superset of B, return True; otherwise, return False.

```
A=set(range(1,11))
B=set(range(5,16))
C=set(range(1,6))
print('A :',A)
print('B :',B)
print(A&B,A.intersection(B),sep=';')
print(A|B,A.union(B),sep=';')
print(A-B,A.difference(B),sep=';')
print(A^B,A.symmetric_difference(B),sep=';')
print(A<=B,A.issubset(B),sep=';')
print(A>=B,A.issuperset(B),sep=';')
print(C<=A,C.issubset(A),sep=';')
```

输出： | Output:

```
A:{1, 2, 3, 4, 5, 6, 7, 8, 9, 10}
B:{5, 6, 7, 8, 9, 10, 11, 12, 13, 14, 15}
{5, 6, 7, 8, 9, 10};{5, 6, 7, 8, 9, 10}
{1, 2, 3, 4, 5, 6, 7, 8, 9, 10, 11, 12, 13, 14, 15};{1, 2, 3, 4, 5, 6, 7, 8, 9, 10, 11, 12, 13, 14, 15}
{1, 2, 3, 4};{1, 2, 3, 4}
{1, 2, 3, 4, 11, 12, 13, 14, 15};{1, 2, 3, 4, 11, 12, 13, 14, 15}
False;False
False;False
True;True
```

6.11 集合拓展示例
6.11 Set Expansion Examples

【例 6-9】假设有一个电子商 | [Example 6-9] Assuming there is an e-commerce

务网站，需要处理大量的用户数据，并找出哪些用户购买了特定的商品，以及这些用户还购买了哪些其他商品。这里可以使用集合来高效地处理这类问题。假设有以下数据：用户列表、商品列表、用户购买记录（每个用户购买的商品列表）。将这些数据表示为 Python 中的集合和字典。

website that needs to handle a large amount of user data. One needs to identify which users have purchased a specific item and what other items these users have also purchased. One can efficiently handle such problems using sets. Assume one has the following data: a list of users, a list of products, and user purchase records(a list of products each user has bought). One can represent these data in Python using sets and dictionaries.

6-9

```
users={'user1', 'user2', 'user3', 'user4', 'user5'}
products={'productA', 'productB', 'productC', 'productD', 'productE'}
purchase_records={
    'user1': {'productA', 'productC', 'productE'},
    'user2': {'productB', 'productD'},
    'user3': {'productA', 'productE'},
    'user4': {'productC', 'productD'},
    'user5': {'productB', 'productE'}
}
target='productA'
set1={user for user, products in purchase_records.items() if target in products}
#Use a set comprehension to identify users who purchased the target product
print(f'Users who purchased {target} are: {set1}')
set2=set()
for user in set1:
    set2.update(purchase_records[user]-{target})
#By iterating through the users in 'set1', add the other products purchased by
#these users(excluding the target product) to 'set2'
print(f'Users who purchased {target} also purchased the following products: {set2}')
t=[y-{target} for x,y in purchase_records.items() if target in y]
#Use a list comprehension to generate a list where each element is the set of
#other products purchased by each user, excluding the target product
print('temp:',t)
if t:
    result=t[0]
    for x in t[1:]:
        set3=result.intersection(x)
else:
    set3=set()
#Based on the contents of 't', identify the set of other products that were
#commonly purchased by users who bought the target product, and store it in 'set3'
print(f'Common items purchased by users who bought {target} : {set3}')
```

185

输出：

```
Users who purchased productA are: {'user3', 'user1'}
Users who purchased productA also purchased the following products: {'productC', 'productE'}
temp: [{'productC', 'productE'}, {'productE'}]
Common items purchased by users who bought productA : {'productE'}
```

Output：

```
Users who purchased productA are: {'user3', 'user1'}
Users who purchased productA also purchased the following products: {'productC', 'productE'}
temp: [{'productC', 'productE'}, {'productE'}]
Common items purchased by users who bought productA : {'productE'}
```

6.12 本章小结
6.12 Summary of This Chapter

本章主要介绍了字典和集合两种数据类型。两种数据类型都可以通过一对大括号"{}"表示。字典元素形式为键值对形式，而集合元素不是键值对形式；字典的键不能重复，集合元素不能重复。具体内容如下：

（1）字典的定义、创建，字典元素的访问、添加、修改和删除，字典的更新与排序；

（2）集合的定义、创建，集合元素的添加和删除，集合的交、并、差等运算。

This chapter mainly introduces two data types: dictionaries and sets. Both data types can be represented using a pair of curly braces "{}". The elements of a dictionary are in key-value pairs, while the elements of a set are not in key-value pairs. Keys in a dictionary must be unique, and elements in a set are unique. Specific content includes:

(1)Definition, creation, accessing dictionary elements, adding, modifying, and deleting elements in a dictionary, updating and sorting dictionaries;

(2)Definition, creation, adding and removing elements in a set, set operations like intersection, union, and set difference.

6.13 本章练习
6.13 Exercise for This Chapter

6.1 设计3个字典dict1、dict2和dict3，每个字典存储一个学生的信息，包括name（姓名）、age（年龄）和id（学号），然后把这3个字典存储到列表student中，遍历这个列表，将其中每个学生的所有信息都输出。

6.2 编写程序，让用户输入一句英文句子，统计每个英文字母出现的次数。

6.3 编写程序，让用户输入一句英文句子，统计每个英文单词出现的次数。

6.1 Design 3 dictionaries dict1, dict2, and dict3, each storing information about a student including name, age, and id. Then store these 3 dictionaries in a list called students, iterate through this list, and output all the information for each student.

6.2 Write a program that prompts the user to input an English sentence, and then counts the occurrences of each English letter.

6.3 Write a program that prompts the user to input an English sentence, and then counts the occurrences of each English word.

6.4 创建一个字典，包含学生的姓名和 Python 成绩数据，计算并输出该字典中所有学生的最高分、最低分和平均分。

6.5 创建两个包含颜色的集合，输出两个集合中共同的颜色。

6.6 创建两个集合，一个集合包含球员 A 喜欢的运动，另一个集合包含球员 B 喜欢的运动，找到并输出两位球员各自独特喜欢的运动和共同喜欢的运动。

6.4 Create a dictionary containing students' names and their Python scores, calculate the highest score, lowest score, and average score among all students in the dictionary, and output the results.

6.5 Create two sets containing colors, output the colors that are common in both sets.

6.6 Create two sets, one containing sports that player A likes and the other containing sports that player B likes, find and output the sports that each player uniquely likes and the sports they both like.

本章练习题参考答案

第 7 章 文件
Chapter 7 File

input()函数可以从键盘接收用户输入的数据，print()函数可以将数据在控制台上显示。这两个标准输入、输出函数主要用于简单的数据处理，一旦程序退出，数据将会丢失。在实际编程中，这些功能往往不能实现更复杂的数据处理和用户交互需求，此时需要借助文件将处理的数据进行保存或者读取。

文件是存储数据的重要形式，包括文本文件和二进制文件两种类型。文本文件存储字符数据，易于阅读和编辑；二进制文件则存储字节数据，适用于图像、音频等。文件操作灵活多样，掌握文件类型、读写、打开、关闭及定位等基本概念，可高效处理各种文件数据。在 Python 编程中，文件操作对数据存储至关重要。

The input() function can receive user input from the keyboard, and the print() function can display data on the console. These two standard input and output functions are mainly used for simple data processing. However, in practical programming, these functions are often not sufficient to meet the needs of more complex data processing and user interaction. In such cases, files can be used to save or read the processed data.

Files are important forms of storing data, including two types: text files and binary files. Text files store character data, which is easy to read and edit; binary files store byte data, suitable for images, audio, etc. File operations are flexible and diverse. Mastering the basic concepts of file types, reading, writing, opening, closing, and positioning can efficiently handle various file data. In Python programming, file operations are crucial for data storage.

7.1 文件概述
7.1 Overview of Files

Python 文件是存储在外部介质，如硬盘、U 盘、云盘等上的数据集合，数据可以是文本、代码、图像等多种形式。文件作为数据的

A Python file is a collection of data stored on external media such as a hard drive, USB drive, cloud storage, etc. The data can take various forms such as text, code, images, etc. Files serve as the carriers of data and play a

第7章 文件
Chapter 7 File

载体，在操作系统中占据重要地位。使用 Python 文件的目的是实现程序与数据的分离，确保数据变动不影响程序的稳定性。同时，文件支持多程序共享与交互，可以提高数据的利用率。

在 Python 中，处理文件时，可以使用相对路径或绝对路径指定文件位置。相对路径是相对于当前工作目录的路径，而绝对路径是从文件系统的根目录开始的完整路径。使用 open()函数打开文件时，需要提供文件路径和使用打开模式。读、写文件通过 read()和 write()等方法进行，之后使用 close()方法关闭文件以释放资源。正确管理文件路径和打开关闭过程是确保文件操作顺利进行的关键。

crucial role in the operating system. The purpose of using Python files is to achieve separation between programs and data, ensuring that changes in data do not affect the stability of the program. Additionally, files support sharing and interaction among multiple programs, thus enhancing the utilization of data.

When working with files in Python, one can specify the file location using either relative paths or absolute paths. A relative path is relative to the current working directory, while an absolute path starts from the root directory of the file system. When using the open() function to open a file, one needs to provide the file path and use the opening mode. Reading and writing to files can be done using methods like read() and write() and then closing the file using close() method to release resources. Properly managing file paths and the open-close process is crucial to ensuring smooth file operations.

7.1.1 文件与文件类型
7.1.1 Files and File Types

文件具有多种属性，这些属性提供了关于文件的详细信息。下面是一些常见的文件属性。

文件名：文件的唯一标识符，通常由主文件名和扩展名组成。

文件类型：由文件的扩展名决定，如.txt 表示文本文件，.py 表示 Python 脚本文件。

文件大小：文件所占用的存储空间大小，通常以字节为单位。

创建时间：文件被创建的日期和时间。

修改时间：文件内容最后一次被修改的日期和时间。

访问时间：文件最后一次被访问的日期和时间。

在 Python（以及更广泛的计算机领域）中，文件被分为文本文件

Files have multiple attributes that provide detailed information about the file. Here are some common file attributes.

File name: The unique identifier of the file, typically composed of the main file name and extension name.

File type: Determined by the file's extension, such as .txt for text files, .py for Python script files name.

File size: The amount of storage space the file occupies, usually measured in bytes.

Creation time: The date and time when the file was created.

Modification time: The date and time when the file's content was last modified.

Access time: The date and time when the file was last accessed.

In Python(as well as the broader computer field), files are primarily categorized into text files and binary

189

和二进制文件主要是基于数据表示和处理的差异。这两种文件类型在存储、读取和处理数据时使用不同的方法和编码。

　　文本文件通常包含字符数据，这些字符是用户可读的，因为它们以标准的字符编码形式（如 UTF-8、ASCII 等）存储。

　　由于字符编码的标准化，文本文件在不同的操作系统和软件之间通常是可移植的。文本文件可以用任何文本编辑器打开和编辑，不需要特定的软件。文本文件常用于存储纯文本信息，如配置文件、日志文件、源代码等。

　　二进制文件以二进制格式存储数据，通常比以文本格式存储的内容更紧凑。二进制格式可以精确地表示原始数据，不会因字符编码转换而丢失信息。这对图像、音频、视频等媒体文件以及特定格式的数据文件（如 Word 文档、Excel 电子表格）至关重要。二进制文件通常需要特定的软件或库来读取和处理，因为它们的格式复杂。对某些类型的数据（如大量数值数据），二进制格式可能提供更快的读写速度和更小的存储空间占用。

files based on the differences in data representation and processing. These two file types use different methods and encodings for storing, reading, and processing data.

Text files typically contain character data that is human-readable because it is stored in standard character encodings format like UTF-8, ASCII, etc.

Due to the standardization of character encodings, text files are usually portable across different operating systems and software. Text files can be opened and edited with any text editor without the need for specific software. Text files are commonly used to store plain text information, such as configuration files, log files, source code, etc.

Binary files store data in a binary format, which is typically more compact than storing content in text format. Binary format can accurately represent raw data without losing information due to character encoding conversion. This is crucial for media files such as images, audio, video, as well as specific data file formats like Word documents, Excel spreadsheets. Binary files usually require specific software or libraries to read and process because their format is complex. For certain types of data, such as large amounts of numerical data, the binary format may offer faster read/write speeds and smaller storage space utilization.

7.1.2 文件存储路径
7.1.2 File Storage Path

　　在计算机系统中，文件可以被存放在不同的位置，这通常被称为文件路径。文件路径不仅包含了文件名，还详细描述了从根目录或当前目录到该文件的具体路径。路径可以分为绝对路径和相对路径两种类型。绝对路径提供从根目录到文件的完整路径，而相对路径则基于

In a computer system, files can be stored in different locations, which is commonly referred to as a file path. The file path not only includes the file name but also details the specific path from the root directory or current directory to the file. Paths can be classified into two types: absolute paths provide the full path from the root directory to the file, while relative paths specify the file's location based on the current working directory.

当前工作目录来指定文件的位置。

在 Windows 操作系统中，绝对路径通常以盘符开始，如 C:\Users\Username\Documents\example.txt。

在 UNIX 或 Linux 系统中，它通常以斜杠（/）开始，如 /home/username/documents/example.txt。使用绝对路径可以确保无论当前工作目录是什么，都能准确地定位到文件。

相对路径是相对于当前工作目录的路径。例如，如果当前工作目录是 C:\Users\Username\Documents，并且有一个名为 example.txt 的文件位于该目录下，那么相对路径就是 example.txt。如果文件位于子目录中，如 C:\Users\Username\Documents\Subfolder\example.txt，则相对路径是 Subfolder\example.txt。

In Windows systems, absolute path typically begins with a drive letter, such as C:\Users\Username\Documents\example.txt.

In UNIX or Linux systems, it usually starts with a forward slash(/), such as /home/username/documents/example.txt. Using an absolute path ensures that the file can be accurately located regardless of the current working directory.

A relative path is a path relative to the current working directory. For example, if the current working directory is C:\Users\Username\Documents and there is a file named example.txt located in that directory, the relative path would be example.txt. If the file is located in a subdirectory, such as C:\Users\Username\Documents\Subfolder\example.txt, then the relative path would be Subfolder\example.txt.

7.1.3 文件处理流程
7.1.3 File Processing Flow

在 Python 中，文件目录是存储文件的组织结构，而文件路径是指向特定文件的地址。处理文件通常包括以下几个步骤。

（1）打开文件：需通过文件路径定位之后进行打开文件操作，可使用内置函数 open() 实现。

（2）读写文件：利用 read()、write() 等方法进行数据的读取和写入。

（3）关闭文件：完成以上操作后，应使用 close() 方法关闭文件，以释放资源。整个过程需确保文件路径正确、打开模式匹配，并妥善处理文件读写异常。

In Python, a file directory is the organizational structure for storing files, while a file path is the address pointing to a specific file. Processing files usually includes the following steps.

(1) Open the file: To open a file after locating it through the file path, one can use the built-in function open().

(2) Read and write to the file: Use methods like read(), write(), etc., for reading and writing data.

(3) Closing the file: After completing the above operations, the file should be closed using the close() method to release resources. Throughout the process, ensure that the file path is correct, the opening mode matches, and handle file reading and writing exceptions properly.

7.2 文件的打开与关闭
7.2 Opening and Closing Files

Python 文件的打开与关闭是文件操作的基础步骤。使用 open() 函数可以打开文件，并返回一个文件对象，通过该文件对象可以进行读写操作。打开文件时需要指定文件名和打开模式，如只读模式 r、写入模式 w 等。为了确保文件被正确关闭并释放资源，通常使用 with 语句来自动管理文件的打开和关闭过程。当 with 语句块执行完毕时，文件会自动关闭，无须手动调用 close()方法。这种做法既简洁又安全，是 Python 中推荐的文件操作方式。

The opening and closing of Python files are fundamental steps in file operations. By using the open() function, one can open a file and obtain a file object, through which one can perform read and write operations. When opening a file, one needs to specify the file name and the opening mode, such as read mode "r" or write mode "w". To ensure that the file is properly closed and resources are released, it is common practice to use the "with" statement to automatically manage the opening and closing process of the file. Once the "with" statement block is executed, the file will be automatically closed without the need to manually call the close() method. This approach is both concise and safe, and is the recommended way of handling file operations in Python.

7.2.1 文件的打开操作
7.2.1 File Opening Operation

在 Python 中，文件的打开操作通常使用内置的 open() 函数来完成。这个函数至少接收两个参数：文件名（如果文件不在当前工作目录下，则应包括路径）和打开模式。打开模式是一个字符串，用于指定文件应该如何被打开和使用。open()函数的语法如下：

In the opening operation of files is typically done using the built-in open() function. This function takes at least two parameters: the file name(including the path if the file is not in the current working directory) and the open mode. The open mode is a string used to specify how the file should be opened and used. Syntax of the open() function is as follows:

```
open(file, encoding=None, mode='r')
```

file 参数：一个字符串，表示要打开的文件名（如果文件不在当前目录下，则应包括路径）。

encoding 参数：用于指定文件字符编码，如UTF-8，当此参数省略时，使用当前操作系统默认的编码类型（Windows 10 一般默认为 GBK 编码，macOS 等一般默认为

The file parameter: A string that represents the file name to be opened(including the path if the file is not in the current directory).

The encoding parameter: Used to specify the character encoding of the file, such as UTF-8, when this parameter is omitted, the default encoding type of the current operating system is used(Windows 10 generally defaults to GBK encoding, macOS generally defaults to

UTF-8 编码)。当使用二进制方式打开文件时,此参数不可使用。

UTF-8 是一种可变长字节表示的 Unicode 字符集编码,可以用 1~6 个字节表示 Unicode 标准中的任何字符。UTF-8 的设计使其向后兼容 ASCII 编码(即 ASCII 字符在 UTF-8 编码中保持不变),并且在处理多字节字符时具有很高的灵活性。因此,在创建文本文件时,建议使用UTF-8编码以方便程序的访问。

mode 参数:一个字符串,表示文件的打开模式。mode 参数设置如表 7-1 所示。

UTF-8 encoding). This parameter cannot be used when opening the file in binary mode.

UTF-8 is a variable-length byte representation of the Unicode character set encoding, which can be used to represent any character in the Unicode standard, and can use 1 to 6 bytes to represent a character. The design of UTF-8 allows it to be backwards compatible with ASCII encoding(i.e. ASCII characters remain unchanged in UTF-8 encoding), and it has a high degree of flexibility in handling multi-byte characters. Therefore, it is recommended to use UTF-8 encoding when creating text files to facilitate program access.

The mode parameter: A string representing the file opening mode. Setting of the mode parameter is as shown in Table 7-1.

表 7-1 mode 参数设置
Table 7-1 Setting of the mode Parameter

符号(Symbol)	含义	Meaning
r	以读模式打开文件(默认值)	Open a file in read mode(default)
w	以写模式打开文件。如果文件已存在,则先清空文件中所有内容;如果文件不存在,则会创建新文件再打开	Open a file in write mode. If the file already exists, clear all content in the file first; if the file does not exist, create a new file and then open it
a	追加模式。如果文件已存在,则数据会被追加到文件所有内容的末尾;如果文件不存在,则会创建新文件再打开	Open a file in append mode. If the file already exists, data will be appended to the end of the existing content; if the file does not exist, create a new file and then open it
x	创建文件并以写模式打开文件。如果文件已存在,则会报错;否则,会创建新文件	Create a file and open it in write mode. If the file already exists, an error will be raised; otherwise, a new file will be created
b	以二进制模式打开文件以处理数据	Open a file in binary mode to process data
+	读写模式(可以与 r、w 或 a 组合)。r+为可读可写,w+为可写可读,a+为可追加写、可读	Read-write mode(can be combined with r, w, or a). r+ for read and write, w+ for write and read, a+ for append and read
t	以文本模式打开文件以处理数据	Open a file in text mode to process data

在 Python 的文件操作中,"+"符号通常与其他的模式字符一起使用,表示同时开启读写功能。有以

In Python file operations, the "+" sign is typically used in conjunction with other mode characters to indicate simultaneous read and write functionality. There

下几种常见的组合模式。

r+：读写模式。文件指针会放在文件的开头，可以读取和写入文件。但是，如果要在读取内容之后写入新的内容，则需要确保文件指针被适当地移动，否则可能会覆盖原有内容。

w+：写读模式。创建一个新文件并进行写入，如果文件已存在则覆盖该文件。写入操作之后，文件指针会放在文件的开头，可以进行读取操作。但是，由于写入时会覆盖文件，所以读取的内容将是空文件或最后一次写入后的内容。

a+：追加和读取模式。文件指针会放在文件的末尾，所以新的内容会被追加到现有内容之后。读取操作需要从文件开头开始，因为追加操作并不会改变读取的位置。

例如，使用 open() 函数打开 my_file1.txt 和 my_file2.txt 文件，文件分别保存在桌面和 D 盘 my_python 文件夹下。

are several common combined modes.

r+: read/write mode. The file pointer is positioned at the beginning of the file, allowing both reading and writing to the file. However, if one intends to write new content after reading, it is important to ensure that the file pointer is appropriately moved to avoid overwriting existing content.

w +: write/read mode. It creates a new file for writing, overwriting the file if it already exists. After a write operation, the file pointer is positioned at the beginning of the file, allowing for reading operations. However, since writing overwrites the file, the content read will either be an empty file or the content from the last write operation.

a+: append and read mode. The file pointer is positioned at the end of the file, allowing new content to be appended after the existing content. Reading operations start from the beginning of the file, as the append operation does not change the reading position.

For example, use the open() function to open two files, my_file1.txt and my_file2.txt, which are saved on the desktop and in the D drive under the my_python folder respectively.

```
#Assuming my_file1.txt is saved on the desktop and my_file2.txt is saved in the folder D: \my_python
#Open the file my_file1.txt using a relative path
>>>f1 = open('my_file1.txt')
#Open the file my_file2.txt using an absolute path
>>>f2 = open('D: /my_python/my_file2.txt')
```

虽然 open() 函数有多个参数，但通常只需要使用前两个参数：file（文件名）和 mode（打开模式）。其他参数在大多数情况下使用默认值就足够了。

在 Python 中，使用 open() 函数会返回一个可迭代对象，可以使用 for 循环遍历的方式访问文件中数据。

Although the open() function has multiple parameters, only the first two parameters are typically needed: file(file name) and mode(opening mode). In most cases, the default values for the other parameters are sufficient.

In Python, using the open() function will return an iterable object that allows one to access the data in a file using a for loop.

Chapter 7　File

[Example 7-1] To read a file with content created on the computer desktop using the open() function.

```
#Open the file for reading
file=open('example.txt', 'r')   #'r' represents read mode
for line in file:
    print(line)
#Close the file
file.close()
```

7-1

Output of traversing text file is shown in Figure 7-1.

Hello, world!

Another line.

Yet another line.

Figure 7-1　Output of Traversing Text File

7.2.2　File Closing Operation

In Python, closing a file is an important step because it releases system resources and ensures that changes to the file are properly saved. After opening a file using the open() function and performing some read or write operations, it is essential to close the file when it is no longer needed. This can be done by calling the close() method on the file object.

However, manually closing a file can sometimes lead to errors, especially in situations like exception handling or forgetting to close the file. Therefore, Python provides a more secure mechanism for automatically closing files, which is using the with statement.

Here are two methods for closing files.

(1) close() method. The close() method does not return a value (it returns None). Its main purpose is to perform the action of closing the file rather than producing a return value. The syntax for close() is as follows:

195

```
file.close()
```

【例7-2】使用open()函数的写数据模式（w）打开文件，并借助write()方法进行"Hello, world！"字符串的写入，最后关闭文件。

[Example 7-2]:Open the file using the write mode (w) with the open() function, then utilize the write() method to write the string "Hello, world！". Finally, close the file.

```
#Open an empty file for writing
file=open('example.txt', 'w')
#Write some data
file.write('Hello, world!')
#Close the file to ensure the write operation is completed and resources are released
file.close()
```

7-2

这个例子中，调用了close()方法来关闭文件。但是，如果在读取文件内容之后发生异常，close()方法可能不会被调用，从而导致资源泄露。

In this example, calling the close() method to close the file. However, if an exception occurs after reading the file content, the close() method may not be called, leading to resource leaks.

（2）with 语句。对于文件操作，with 语句可以确保文件在使用后被正确关闭，即使在处理文件时发生异常也是如此。with 语句的语法如下：

(2)The with statement. For file operations, the with statement ensures that the file is correctly closed after its use, even in the presence of exceptions during file handling. The syntax of the with statement is as follows:

```
with expression as variable:
    #Perform operations using the variable within this block
    #When the block is completed, the __exit__() method of expression will be called
```

对于文件操作，expression 通常是 open()函数，它会返回一个文件对象。这个文件对象会被赋值给 variable 变量，可以在随后的代码块中使用这个变量来引用文件。当 with 语句结束时，文件对象的__exit__()方法会被自动调用，从而关闭文件。

For file operations, the expression is typically the open() function, which returns a file object. This file object is assigned to a variable and can be used to reference the file in subsequent code blocks. When the with statement ends, the file object's __exit__() method is automatically called to close the file.

【例7-3】下面是一个使用with语句打开和关闭文件的例子。

[Example 7-3] Here is an example of opening and closing a file using the with statement.

第7章 文件
Chapter 7　File

```
#Open the file using the with statement
with open('example.txt', 'r') as file:
    #Read the contents of the file
    content=file.read()
    #Handle the file content, for example, print it to the console
    print(content)
#The file has been automatically closed, so there is no need to call file. close().
```

7-3

在这个例子中，open('example.txt', 'r')是expression，它返回一个文件对象，并且该文件对象被赋值给变量file。

在with语句的代码块内，可以使用file变量来读取文件内容。当代码块执行完毕时，文件会被自动关闭，无须显式调用file.close()方法。

即使在with语句内部发生异常时，文件仍然会被正确关闭。这是因为with语句确保了上下文管理器的__exit__()方法会被调用，而文件对象的上下文管理器会在该方法中关闭文件。

In this example, open('example.txt', 'r') is the expression, which returns a file object. This file object is assigned to the variable file.

Inside the code block of the with statement, the variable "file" can be used to read the file contents. When the code block completes execution, the file is automatically closed, and there is no need to explicitly call file.close().

If an exception occurs within the with statement, the file will still be closed correctly. This is ensured by the with statement guaranteeing that the __exit__() method of the context manager is called, and the file object's context manager will close the file within that method.

7.2.3　文件的基本操作实例
7.2.3　Basic File Operation Examples

下面是读取文件模式的用法。

The following is the usage of the file reading mock.

```
#Open the file in read-only mode
with open('example.txt', 'r') as file:
    content=file.read()
```

在使用只读模式打开文件之前需要先创建example.txt文件，否则会出现以下错误提示：

Before executing the file opening in read mode, one needs to create the example.txt file; otherwise, an error message will be displayed:

```
Traceback(most recent call last):
    File "C:/Users/Administrator/Desktop/test.py", line 2, in<module>
        with open('example.txt', 'r') as file:
FileNotFoundError: [Errno 2] No such file or directory: 'example.txt'
```

197

可以先在桌面上创建 example.txt，这样执行时就不会出现问题，读操作就能成功运行。接下来使用写操作来对空文件进行内容的更改。

The example.txt file can be created on the desktop first, so the execution will proceed without any issues, and the read operation will run successfully. Next, use the write operation to modify the content of the empty file.

```
#Open the file in write mode, existing content will be overwritten
with open('example.txt', 'w') as file:
    file.write('Hello, world!')
```

运行程序后，会在 example.txt 文档中显示"Hello, world!"这串字符，如图 7-2 所示。

After running the program, the text "Hello, world!" will be displayed in the example.txt document, and the execution result is shown as in Figure 7-2.

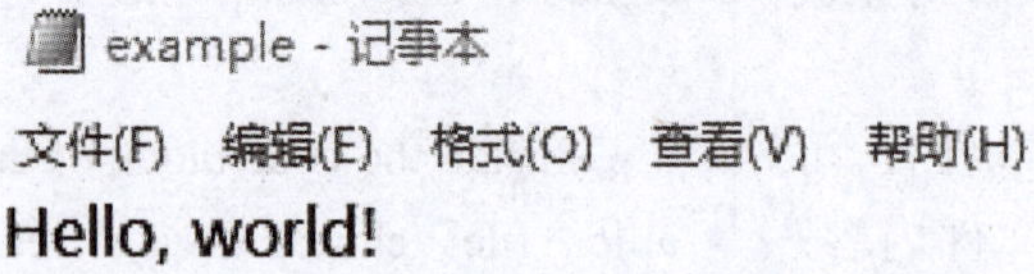

图 7-2　写操作执行结果
Figure 7-2　Result of the Write Operation

使用"a"模式打开一个文件时，任何写入操作都会将数据追加到文件的末尾，而不会覆盖现有的内容。如果文件不存在，则会创建一个新文件。这是其与"w"模式的主要区别，"w"模式会覆盖文件中的任何现有内容。以下是一个使用"a"模式打开文件的简单示例。

When opening a file in "a" mode, any write operation will append data to the end of the file without overwriting existing content. If the file does not exist, a new file will be created. This is the main difference compared to the "w" mode, which overwrites any existing content in the file. Here is a simple example of opening a file in "a" mode.

```
#Open the file in "a" mode, new content will be added to the end of the file.
with open('example.txt', 'a') as file:
    file.write('\nAnother line. ')
```

程序运行之后将新的字符串写入了 example.txt 文件末尾，而不覆盖其原有内容，如图 7-3 所示。

The new string was written to the end of the example.txt file after the program ran, without overwriting its original content, as shown in Figure 7-3.

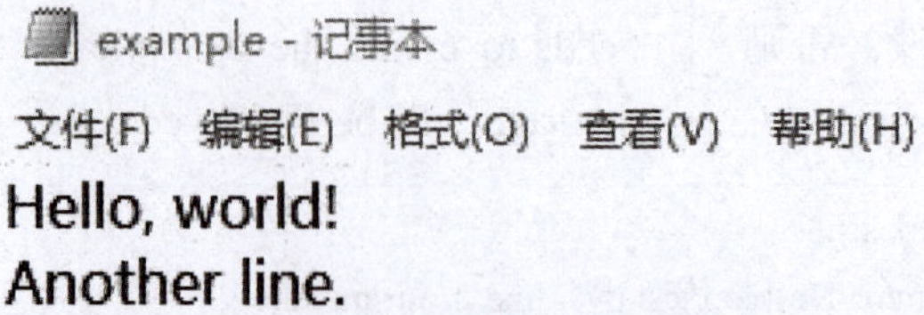

图 7-3　追加操作执行结果
Figure 7-3　Result of the Append Operation

第7章 文件
Chapter 7 File

在 Python 中，使用"r+"模式打开文件表示以读写模式打开文件。这意味着可以读取文件的内容，也可以向文件中写入内容。然而，与纯粹的写入模式（"w"）不同，"r+"模式不会截断文件（即不会删除文件的现有内容）。

与追加模式（"a"）也不同，"r+"模式允许在文件的任何位置写入内容，而不仅仅是在文件末尾。

"a+"模式用于打开一个文件以进行追加和读取。如果文件不存在，则文件将被创建。如果文件已经存在，则数据将被追加到文件的末尾，而不会覆盖现有的内容。此外，可以读取文件的内容。

【例 7-4】如果 example.txt 文本文件中，有一行字符串"Hello, world!"，此时用户想在文件末尾追加一个字符串"This is appended text.\n"，并实现读取操作。

In Python, using the "r+" mode to open a file means opening the file in read-write mode. This allows one to read the file's content and also write to the file. However, unlike the write-only mode（"w"）, the "r+" mode does not truncate the file（i. e., it does not delete the existing content of the file）.

Unlike the append mode（"a"）, the "r+" mode allows writing at any position in the file, not just at the end of the file.

"a+" mode is used in Python to open a file for both appending and reading. If the file does not exist, it will be created. If the file already exists, data will be appended to the end of the file without overwriting the existing content. Additionally, it allows reading the content of the file.

[Example 7-4] If the text file example.txt contains the line "Hello, world!", and the user wants to append the string "This is appended text. \n" to the end of the file and perform a read operation.

```
#To open a file for append and read operations
with open('example.txt', 'a+') as file:
    #Append some text to the end of the file
    file.write('This is appended text.\n')
    #Move the file pointer to the beginning of the file for reading
    file.seek(0)
    #Read and print the file content
    content = file.read()
    print(content)
```

7-4

输出：

Output:

Hello,world!This is appended text.

"w+"模式用于打开一个文件以进行写入和读取操作。如果文件不存在，则文件将被创建。如果文件已经存在，则它的内容将被覆盖。写入操作之后，可以重新定位文件指针来读取文件的内容。

"w+" mode is used to open a file for both writing and reading operations. If the file does not exist, it will be created. If the file already exists, its content will be overwritten. After the write operation, the file pointer can be repositioned to read the content of the file.

【例 7-5】如果 example.txt 文本文件中，有一行字符串"Hello, world!"，此时用户想覆盖写入一个字符串"This is new content.\n"，并实现读取操作。

[Example 7-5] If the text file example.txt contains the line "Hello, world!", and the user wants to write the string "This is new content. \n" to overwrite the file and perform a read operation.

```
#Open a file for writing and reading
with open('example.txt', 'w+') as file:
    #Write some text(this will overwrite the original content in the file)
    file.write('This is new content. \n')
    #Move the file pointer to the beginning of the file for reading
    file.seek(0)
    #Read and print the file content
    content = file.read()
    print(content)
```

7-5

输出：

Output:

```
This is new content.
```

7.3 文件的基本操作
7.3 Basic Operations of Files

Python 文件的基本操作除打开和关闭外，还包括读取、写入、定位、重命名和删除等。在这些操作中，读取和写入操作是非常重要的。

In addition to opening and closing files, the basic operations for Python files include reading, writing, positioning, renaming, and deleting. Among these operations, reading and writing operations are particularly important.

7.3.1 文件的读写操作
7.3.1 Reading and Writing Operations on Files

文件的读写操作是编程中的重要功能。在 Python 中，通常使用 open()函数打开文件，并通过文件对象的方法如 read()、readline()、readlines()进行读取。read()用于读取整个文件或指定数量的字符，readline()读取一行，readlines()则读取所有行并返回列表。写入文件时，使用 write()或 writelines()方法，前者写入一个字符串，后者写入字

Reading and writing operations on files are essential functions in programming. In Python, the open() function is typically used to open a file, and reading can be done using methods like read(), readline(), and readlines() on the file object. The read() is used to read the entire file or a specified number of characters, readline() reads one line, and readlines() reads all lines and returns a list. When writing to a file, the write() or writelines() methods can be used, with the former writing a string and the latter writing a list of strings. It is important to close the

第7章 文件
Chapter 7 File

符串列表。操作完成后需关闭文件以释放资源。

（1）读取文件。

①read()：读取文件的全部内容。语法如下：

```
file.read(size)
```

size 是一个可选的数字参数（可省略）。当指定了 size 时，读取指定数量的数据，然后返回字符串。当未指定 size 或指定其为负数时，读取并返回整个文件。

【例 7-6】在 example.txt 文件中，已有内容为一个字符串"This is new content."，使用 read()方法读取整个文件内容或读取前 10 个字符。

file after operations to release resources.

(1)Read the File.

①read(): Read the entire contents of the file. The syntax is as follows:

```
file.read(size)
```

The "size" is an optional numerical parameter(can be omitted). When "size" is specified, it reads the specified amount of data and then returns it as a string. If "size" is not specified or is specified as a negative number, it reads and returns the entire file.

[Example 7-6] The example.txt file already contains the string "This is new content.", one can read the entire file content or the first 10 characters using the read() method.

```
with open('example.txt', 'r') as file:
    content=file.read()          #Read the entire content of the file
    print(content)
with open('example.txt', 'r') as file:
    first_10_chars=file.read(10)  #Read the first 10 characters
    print(first_10_chars)
```

7-6

输出：

```
This is new content.
This is ne
```

Output:

```
This is new content.
This is ne
```

②readline()：读取文件的一行内容。换行符\n 留在字符串末尾，只有在到达文件末尾时才返回一个空字符串。语法如下：

```
file.readline()
```

【例 7-7】在 example.txt 文件中，已有两行字符串，第一行字符串是"This is the first line."，第二行字符串是"This is the second line."，如图 7-4 所示，用 readline()方法来

②readline(): Read a single line from the file. The newline character \n is retained at the end of the string, and only returns an empty string when it reaches the end of the file. The syntax is as follows:

```
file.readline()
```

[Example 7-7] In the file example.txt, there are two lines of strings. The first line contains "This is the first line." and the second line contains "This is the second line." As shown in Figure 7-4, using the readline() method to read a line from the file, the reference code is as fol-

201

读取文件中的一行,参考代码如下,运行结果如图 7-5 所示。

lows, and the execution result is shown in Figure 7-5.

图 7-4 example.txt 文件内容

Figure 7-4 Content of the example.txt File

```
with open('example.txt', 'r') as file:
    line=file.readline()        #Read the first line of content
    print(line)
```

================== RESTART: C:\Users\Administrator\Desktop\test.py ==
This is the first line.

图 7-5 运行结果

Figure 7-5 Execution Result

如果成功读取了一行,readline()方法将返回一个字符串,其中包含该行的内容(包括行尾的换行符)。如果已经到达文件的末尾,或者文件是空的,readline()方法将返回一个空字符串("")。

If readline() method successfully reads a line, it will return a string containing the content of that line (including the newline character at the end of the line). If the end of the file has been reached or the file is empty, readline() method will return an empty string("").

③readlines():读取整个文件所有行并返回一个包含各行(作为元素)的列表。每一行后面的换行符 \n 都会保留。语法:

③readlines(): Read all lines of the entire file and return a list containing each line as an element. The newline character \n at the end of each line is retained. Syntax:

```
file.readlines()
```

【例 7-8】在 example.txt 文件中,已有三行字符串,第一行字符串是"This is the first line.",第二行字符串是"This is the second line.",第三行字符串是"This is the third line.",如图 7-6 所示。使用 readlines()方法来读取文件中的每一行,参考代码如下,运行结果如图 7-7 所示。

[Example 7-8] In the file example.txt, there are three lines of strings. The first line contains "This is the first line.", the second line contains "This is the second line.", and the third line contains "This is the third line.", as shown in Figure 7-6. Using the readlines() method to read each line from the file, the reference code is as follows, and the execution result is shown in Figure 7-7.

第7章 文件
Chapter 7　File

*example.txt - 记事本
文件(F)　编辑(E)　格式(O)　查看(V)　帮助(H)
This is the first line.
This is the second line.
This is the third line.

图 7-6　example.txt 文件内容

Figure 7-6　Content of the example.txt File

```
with open('example.txt', 'r') as file:
    lines = file.readlines()   #Read all lines into a list
    for line in lines:
        print(line, end='')
```

================== RESTART: C:\Users\
This is the first line.
This is the second line.
This is the third line.

图 7-7　运行结果

Figure 7-7　Execution Result

（2）文件的写操作。

①write(string)：将 string 写入文件，返回写入的字符数。如果要在文件中写入字符串以外的数据，则需要先将数据转换为字符串。语法：

(2)File writing operation.

①write(string): Write the string to the file and return the number of characters written. If one wants to write data other than strings to a file, one needs to convert the data to a string first. Syntax:

```
file.write(string)
```

【例 7-9】使用 open() 函数打开一个名为 example_write.txt 的文件，如果该文件不存在，Python 将会创建它。w 表示写模式，这意味着如果文件已经存在并且其中有内容，那么该文件的内容将被覆盖。

[Example 7-9] Open a file named example_write.txt using the open() function. If the file does not exist, Python will create it. "w" represents write mode, which means that if the file already exists and contains content, the content of the file will be overwritten.

```
with open('example_write.txt', 'w') as file:    #The file example_write.txt is empty.
    num_chars = file.write('Hello, world!')
    print(f"Number of characters written: {num_chars}")
```

"with…as file:"是一个上下文管理器，它将打开的文件对象赋值给变量 file。可以通过 file 这个变

"with…as file:" is a context manager that assigns the opened file object to the variable "file". One can reference and operate on the file using this variable "file".

203

量来引用和操作该文件。with 语句的好处是它会在代码块结束后自动关闭文件。

"num_chars = file.write('Hello, world!')"这行代码调用 file 对象的 write()方法,将字符串"Hello, world!"写入文件。write()方法返回写入的字符数,这个返回值被赋值给变量 num_chars。运行结果如下:

The benefit of the with statement is that it automatically closes the file after the code block ends.

The code "num_chars = file.write('Hello, world!')" calls the write() method of the file object to write the string "Hello, world!" to the file. The write() method returns the number of characters written, and this return value is assigned to the variable num_chars. The output is as follows:

```
#Output
Number of characters written: 13
```

②writelines(lines):向文件写入一个字符串列表(lines),不自动添加换行符。通常,lines 中的每个字符串都以换行符\n结尾。语法:

②writelines(lines): Write a list of strings(lines) to the file without adding automatic line breaks. Typically, each string in lines ends with a newline character "\n". Syntax:

```
file.writelines(lines)
```

【例 7-10】writelines(lines)函数的使用。

[Example 7-10] Usage of the writelines(lines) function.

```
lines = ['First line\n', 'Second line\n', 'Third line\n']
with open('example_writelines.txt', 'w') as file:
    file.writelines(lines)    #Write the string list
```

7-10

"lines = ['First line\n', 'Second line\n', 'Third line\n']"这行代码创建了一个名为 lines 的列表,其中包含 3 个字符串元素。每个字符串都以换行符\n 结尾,当这些字符串被写入文件时,每个字符串后面都会有一个换行。

"file.writelines(lines)"这行代码调用 file 对象的 writelines()方法,将整个 lines 列表中的字符串一次性写入文件中。与 write ()方法不同,writelines()方法不会在每个字符串之间自动添加换行符,所以每

The line of code "lines = ['First line\n', 'Second line\n', 'Third line\n']" creates a list named "lines" that contains three string elements. Each string ends with a newline character "\n", so when these strings are written to a file, there will be a new line after each string.

The code "file.writelines(lines)" calls the "writelines()" method of the file object to write all the strings from the "lines" list to the file at once. Unlike the write() method, writelines() method does not automatically add newline characters between each string, so each string must already include the necessary newline character(In this

个字符串中必须已经包含了所需的换行符(在本例中,每个字符串都以\n结尾)。运行结果在 example_writelines.txt 文件中显示如下:

case, each string ends with " \n"). The output in the "example_writelines.txt" text file will appear as follows:

First line
Second line
Third line

7.3.2 文件的定位操作
7.3.2 File Positioning Operations

在 Python 中,seek()和 tell()是文件对象(通常是使用 open()函数打开的文件)的两种方法,它们用于处理文件的读取/写入位置。

(1)seek(offset, whence = 0)方法用于移动文件的读取/写入指针到指定的位置。这个方法有两个参数:offset 和 whence。

offset:偏移量,表示要移动的字节数。

whence(可选):基准点,默认是 0。它有 3 个可能的值:0、1、2。

①0:文件的起始位置。这是默认值。此时 offset 必须是非负的。

②1:文件的当前位置。此时 offset 可以是负的。

③2:文件的结束位置。此时 offset 通常是负的,但也可以是 0 或正数(这通常没有多大意义,因为这会超出文件的当前大小)。

【例 7-11】seek()方法的使用。

In Python, seek() and tell() are two methods of file objects(typically files opened using the open() function) used to handle the reading/writing position within a file.

(1)The seek(offset, whence = 0) method is used to move the file's read/write pointer to a specified position. This method has two parameters: offset and whence.

offset: The offset parameter represents the number of bytes to move.

whence(optional): The reference point, which is 0 by default. It has three possible values: 0,1,2.

①0: The beginning of the file. This is the default value. The offset must be non-negative.

②1: The current file position. The offset can be negative.

③2: The end of the file. The offset is usually negative but can also be 0 or positive(although this often makes little sense as it extends beyond the current size of the file).

[Example 7-11] Usage of the seek() function.

```
#Create a simple empty text file as an example
with open('example_seek.txt', 'w') as file:
    file.write('Hello, world! \nThis is a test. \nEnd of file. ')
#Open the file to read the text content
with open('example_seek.txt', 'r') as file:
    #Read the first 5 characters
    print(file.read(5))          #Output: Hello
```

7-11

```
#Use seek() to move the file pointer to the beginning of the file
file.seek(0)
#Read the first 5 characters again to confirm that seek() has taken effect
print(file.read(5))          #Output: Hello
#Use seek() to move the file pointer to the 7th byte position
#In ASCII encoding, each character occupies 1 byte, so it actually moves to after the 7th character file.seek(7)
#Read the characters until the end of the line
print(file.readline())       #Output: world!, and a newline '\n' followed by a blank line
#Move to the end of the file and read
file.seek(0, 0)              #Reset the file read pointer to the beginning, with offset 0 and whence 0
print(file.read())           #Output: ''(empty string)
```

在这个例子中,首先创建一个包含 3 行文本的简单文件。然后,打开这个文件进行读取操作,并使用 seek()来移动文件指针到不同的位置。同时使用了 read()来读取文件内容。

example_seek.txt 文本文件中包含 3 行字符串,如下所示。

In this example, first create a simple file containing three lines of text. Then, open the file for reading operation, and use seek() to move the file pointer to different positions. The read() function is also used to read the file contents.

The text file example_seek.txt contains three lines of strings as follows.

```
Hello, world!
This is a test.
End of file.
```

运行后的结果如下所示。

The result after running is as follows.

```
Hello
Hello
world!
Hello, world!
This is a test.
End of file.
```

(2)tell()方法在 Python 的文件操作中用于获取当前文件读取或写入指针的位置。这个方法返回的是从文件开头到当前指针位置的字节数。

【例 7-12】下面是一个使用 tell()方法的简单例子,在读取文本文件时跟踪当前读取指针位置。

(2)The tell() method in Python's file operations is used to obtain the current position of the file read or write pointer. This method returns the number of bytes from the beginning of the file to the current pointer position.

[Example 7-12] Here is a simple example using the tell() method to track the current read pointer position when reading a text file.

Chapter 7　File

```python
#First, create a text file and write content to it
with open('example_tell.txt', 'w', encoding='utf-8') as file:
    file.write('Hello, this is a test text file. \nWe will use tell() to find the current position. ')
#Open the file and read the contents, while using tell() to check the position
with open('example_tell.txt', 'r', encoding='utf-8') as file:
    #Read the first 5 characters
    print(file.read(5))                       #Output: Hello
    #Check the current file pointer position
    position=file.tell()
    print(f"Current file position: {position}")   #Output: Current file position: 5
```

7-12

example_tell.txt 文本文件中包含两行字符串，如下所示。 | The text file example_tell.txt contains two lines of strings as follows.

Hello, this is a test text file.
We will use tell() to find the current position.

运行后的结果如下所示。 | The result after running is as follows.

Hello
Current file position: 5

在这个例子中，首先打开文件并写入了 2 行文本。然后，重新打开文件以进行读取，并使用 read() 方法读取了前 5 个字符。之后，使用 tell() 方法查看了当前的文件指针位置，它应该是 5，因为读取了 5 个字符。

In this example, the file is first opened and two lines of text are written. Next, the file is reopened for reading, and the read () method is used to read the first 5 characters. Then, the tell() method is used to check the current file pointer position, which should be 5 because 5 characters were read.

7.3.3　文件的其他操作
7.3.3　Other Operations on Files

（1）f.flush()：刷新文件内部缓冲区。在写入文件时，数据通常首先被写入一个内部缓冲区，然后在适当的时候（如缓冲区满或文件关闭时）才写入磁盘。flush()方法强制将缓冲区中的数据立即写入磁盘。参数：无。返回值：无。

（2）f.truncate(size=None)：调整文件的大小。如果 size 参数小于文件的当前大小，则文件将被截

(1) f.flush(): Flush the internal buffer of the file. When writing to a file, data is typically first written to an internal buffer and then written to disk at the appropriate time(such as when the buffer is full or the file is closed). The flush() method forces the data in the buffer to be immediately written to disk. Parameters: None. Return value: None.

(2) f.truncate(size=None): Adjust the size of the file. If the size parameter is smaller than the current size of the file, the file will be truncated, and the excess data will be

断，多余的数据会丢失。如果 size 参数大于或等于文件的当前大小，则文件将使用空字节进行填充（在某些系统上可能不支持填充）。参数：size，新的文件大小，以字节为单位。如果省略该参数或该参数为 None，则文件将被截断为 0 字节。返回值：无。

（3）f.closed：是一个属性，不是一种方法。返回一个布尔值，表示文件是否已关闭。返回值：如果文件已关闭，则返回 True；否则返回 False。

（4）f.fileno()：返回一个整数，表示文件的底层文件描述符。参数：无。返回值：文件的底层文件描述符。

（5）f.readable()：检查文件是否可读。参数：无。返回值：如果文件可读，则返回 True；否则返回 False。

（6）f.writable()：检查文件是否可写。参数：无。返回值：如果文件可写，则返回 True；否则返回 False。

（7）f.seekable()：检查文件是否支持随机访问。随机访问是指，可以使用 seek()方法跳转到文件的任意位置。参数：无。返回值：如果文件支持随机访问，则返回 True；否则返回 False。

（8）f.isatty()：检查文件是否连接到一个终端设备（如控制台、终端窗口等）。参数：无。返回值：如果文件连接到一个终端设备，则返回 True；否则返回 False。

这些方法和属性提供了对文件对象进行各种操作和查询的能力。在使用它们时，请确保文件对象已

lost. If the size parameter is greater than or equal to the current size of the file, the file will be padded with null bytes(padding may not be supported on some systems). Parameters: size, the new size of the file in bytes. If omitted or set to None, the file will be truncated to 0 bytes. Return value: None.

(3)f.closed: This is an attribute, not a method. Return a boolean value indicating whether the file is closed. Return value: True if the file is closed, False otherwise.

(4)f.fileno(): Return an integer representing the underlying file descriptor of the file. Parameters: None. Return value: The underlying file descriptor of the file.

(5)f.readable(): Check if the file is readable. Parameters: None. Return value: True if the file is readable, False otherwise.

(6)f.writable(): Check if the file is writable. Parameters: None. Return value: Return True if the file is writable, False otherwise.

(7)f.seekable(): Check if the file supports random access, meaning that the seek() method can be used to navigate to any position in the file. Parameters: None. Return value: Return True if the file supports random access, False otherwise.

(8)f.isatty(): Check if the file is connected to a terminal device(such as a console or terminal window). Parameters: None. Return value: True if the file is connected to a terminal device, False otherwise.

These methods and attributes provide the ability to perform various operations and queries on file objects. When using them, make sure the file object is properly

7.4 文件的综合应用
7.4 Integrated Application of Files

Python 提供了丰富的文件操作功能，包括读取、写入、追加、删除、移动等。下面将展示两个综合应用案例，展示如何使用 Python 文件方法进行文件操作。

Python provides a rich set of file operation functions, including reading, writing, appending, deleting, moving files, and more. Below, two comprehensive application examples will be demonstrated to show how to use Python file methods for file operations.

【例 7-13】文本文件内容统计。假设有一个文本文件 example_case1.txt，里面包含了一些文本内容，如下所示。任务是统计文件中每个单词出现的次数。

[Example 7-13] Text File Content Statistics. Assuming there is a text file named example_case1.txt that contains some text content as shown below. The task is to count the occurrences of each word in the file.

Hello this is a sample text file
It contains some words that we want to count
Each word in this text file will be counted
Words like 'the' 'a' and 'is' are common words
But we will count every word, regardless of how common it is
Even words like 'file' and 'text' will be counted
Note that punctuation marks like commas and periods are not counted as words
However, if words are separated by punctuation, they will still be counted as separate words.

7-13

算法分析如下。

(1) 首先使用 with 语句来打开名为 example_case1.txt 的文件。with 语句可以确保文件在操作完成后被正确关闭。文件被以只读模式(r)打开，并指定 UTF-8 编码，这是处理包含非 ASCII 字符的文本文件时的常见做法。使用 file.read()方法读取文件的全部内容，并将其存储在变量 content 中。参考代码如下：

The algorithm analysis is as follows.

(1)Firstly, use the "with" statement to open the file named example_case1.txt. The "with" statement ensures that the file is correctly closed after the operation is completed. The file is opened in read-only mode(r) and specifies the UTF-8 encoding, which is a common practice for handling text files containing non-ASCII characters. Use the file.read() method to read the entire content of the file and store it in the variable "content". The reference code is as follows:

```
#Open the file and read the content
with open('example_case1.txt', 'r', encoding='utf-8') as file:
    content=file.read()
```

209

（2）接下来，使用 content.split() 方法将文件内容分割成一个单词列表 words。默认情况下，split()方法会根据任意数量的空白字符（空格、换行符、制表符等）来分割字符串。它会把连续的空白字符当作一个分隔符来处理。

如果文件内容包含其他分隔符（如逗号、冒号等），或者需要考虑单词的大小写、标点符号等问题，那么就需要更复杂的分割逻辑。参考代码如下：

(2)Next, the content.split() method is used to split the file content into a word list "words". By default, the split() method will split the string based on any number of whitespace characters(spaces, newline characters, tabs, etc.). It treats consecutive whitespace characters as a single separator.

If the file content contains other separators(such as commas, colons, etc.), or if one needs to consider issues like word case sensitivity, punctuation, etc., then a more complex splitting logic will be required. Reference code is as follows:

```
words=content.split()
```

（3）创建一个空字典 word_count 来存储每个单词及其出现的次数。接下来，通过 for 循环遍历单词列表 words。对列表中的每个单词，检查它是否已经在字典 word_count 中。如果是，则将其计数增加 1；如果不是，则将其添加到字典中，并设置计数为 1。参考代码如下：

(3)Create an empty dictionary word_count to store each word and its frequency of occurrence. Then, iterate through the word list "words" using a for loop. For each word in the list, the code checks if it is already in the dictionary "word_count". If it is, increment its count by 1; if it is not, add it to the dictionary and set the count to 1. Reference code is as follows:

```
#Use a dictionary to count the occurrences of each word
word_count={}
for word in words:
    if word in word_count:
        word_count[word]+=1
    else:
        word_count[word]=1
```

（4）最后，使用 for 循环遍历字典 word_count 中的键值对，并打印出每个单词及其出现的次数。参考代码如下：

(4)Finally, iterate through the key-value pairs of the dictionary "word_count" using a for loop and print out each word along with its frequency of occurrence. Reference code is as follows.

```
#Print the results
for word, count in word_count.items():
    print(f"{word}: {count}")
```

第7章 文件
Chapter 7 File

例 7-13 的完整参考代码如下：

The complete reference code for Example 7-13 is as follows:

```
#Open the file and read the content
with open('example_case1.txt', 'r', encoding='utf-8') as file:
    content=file.read()
#Split the content into a word list
#Here, we simply use a space as the delimiter, but in reality, more complex splitting logic
#may be required
words=content.split()
#Use a dictionary to count the occurrences of each word
word_count={}
for word in words:
    if word in word_count:
        word_count[word]+=1
    else:
        word_count[word]=1
#Print the results
for word, count in word_count.items():
    print(f"{word}: {count}")
```

总的来说，这段代码是一个简单的文本文件分析工具，用于统计文本文件中每个单词的出现次数。它展示了 Python 中文件操作、字符串处理和字典使用的基础知识。

In general, this piece of code is a simple text file analysis tool used to count the occurrences of each word in a text file. It demonstrates the basic knowledge of file operations, string handling, and dictionary usage in Python.

【例 7-14】日志文件分析。假设有一个日志文件 log.txt，每行记录了一个访问请求，格式如下。任务是找出所有 ERROR 级别的日志，并将它们写入一个新的文件 errors.txt。

[Example 7-14] Log file analysis. Suppose there is a log file named log.txt, where each line records a visit request in the following format. The task is to identify all logs at the ERROR level and write them to a new file named errors.txt.

```
2024-03-06 12:00:01,INFO,Module,This is some info.
2024-03-06 12:01:30,ERROR,Module,This happened!
```

算法分析如下。

(1)打开文件。使用 with 语句同时打开两个文件：log.txt（只读模式，编码为 UTF-8）和 errors.txt（写入模式，编码为 UTF-8）。

with 语句的好处是，当代码块

The algorithm analysis is as follows.

(1)Open the files. Use a with statement to open two files simultaneously: log.txt (in read mode, encoded as UTF-8) and errors.txt (in write mode, encoded as UTF-8).

The benefit of using a with statement is that the files

211

执行完毕后，文件会被自动关闭，无须显式调用 close()方法。log.txt 被打开用于读取，其内容可能是某种日志信息。errors.txt 被打开用于写入，将用于存储从 log.txt 中筛选出的错误信息。

（2）逐行读取和处理：通过 for 循环逐行读取 log.txt 中的内容。对于每一行，检查是否包含字符串 ERROR。这是一种简单的日志级别检查方式，假设日志行中包含 ERROR，则表示这是一个错误日志。如果当前行包含 ERROR，则将该行写入 errors.txt 文件中。完整参考代码如下：

will be automatically closed once the code block is executed, without the need to explicitly call the close() method. log.txt is opened for reading, containing some form of log information. errors.txt is opened for writing, to store the error information filtered from log.txt.

(2)Read and process line by line: Iterate through the content of log.txt line by line using a for loop. For each line, check if it contains the string ERROR. This is a simple way of checking the log level, assuming that a line containing ERROR indicates an error log. If the current line contains ERROR, write that line to the errors.txt file. Complete reference code is as follows:

```python
#Open the log file and the new file
with open('log.txt', 'r', encoding='utf-8') as log_file, open('errors.txt', 'w', encoding='utf-8') as error_file:
    #Read the log file line by line
    for line in log_file:
        #Check if the log level is ERROR
        if 'ERROR' in line:
            #If it is ERROR, then write it to the new file
            error_file.write(line)
```

7-14

这个案例展示了如何使用 Python 读取和处理文件，以及如何根据特定条件筛选和写入文件内容。注意，在实际应用中，日志文件的格式可能更加复杂，需要更复杂的解析和处理逻辑。但是，上面的代码提供了一个基本的框架和思路。

This case demonstrates how to read and process files using Python, as well as how to filter and write file contents based on specific conditions. It is important to note that in actual applications, the format of log files may be more complex, requiring more advanced parsing and processing logic. However, the code provided above offers a basic framework and approach.

7.5 本章小结
7.5 Summary of This Chapter

（1）文件类型与打开模式。在 Python 中，文件可以通过内置函数 open()打开，并指定文件类型和打开模式。常见的文件类型包括文本

(1)File types and opening modes. In Python, files can be opened using the built-in function open(), specifying the file type and opening mode. Common file types include text files and binary files.

文件和二进制文件。

打开模式如 r 表示只读，w 表示写入（会覆盖原有内容），a 表示追加，以及 b 表示二进制模式。例如，rb 或 wb 用于读写二进制文件。

（2）存储路径。文件的存储路径可以是绝对路径或相对路径。绝对路径从文件系统的根目录开始，而相对路径是相对于当前工作目录而言。在跨平台应用中，建议使用 os.path 模块来处理文件路径，以确保代码的可移植性。

（3）文件打开与关闭。使用 with 语句可以自动管理文件的打开和关闭，这是一种推荐的做法，因为它能确保即使在发生异常时文件也能被正确关闭。如果不使用 with 语句，则需要显式地调用 close() 方法来关闭文件。

（4）文件读写。对于文本文件，可以使用 read()、readline() 和 readlines() 方法来读取内容，使用 write() 方法来写入内容。对于二进制文件，可以使用类似的 read() 和 write() 方法，但需要以二进制模式打开文件。

（5）文件定位。seek(offset, whence) 方法用于移动文件指针到指定位置。offset 表示偏移量，whence 指定起始位置（0 表示文件开头，1 表示当前位置，2 表示文件结尾）。tell() 方法返回当前文件指针的位置。

（6）文件其他操作。flush() 方法用于将缓冲区的数据立即写入文件，而不用等待缓冲区满或文件关闭。

truncate(size) 方法可以调整文件的大小。如果指定的 size 比文件

Opening modes like "r" for read, "w" for write (which will overwrite existing content), "a" for append, and "b" for binary mode. For example, "rb" or "wb" is used for reading and writing binary files.

(2)Storage paths. File storage paths can be absolute paths or relative paths. An absolute path starts from the root directory of the file system, while a relative path is relative to the current working directory. In cross-platform applications, it is recommended to use the os.path module to handle file paths to ensure code portability.

(3) Opening and closing files. Using the with statement can automatically manage the opening and closing of files, which is a recommended practice as it ensures that files are correctly closed even in the event of an exception. If the with statement is not used, the close() method needs to be explicitly called to close the file.

(4)File reading and writing. For text files, one can use the read(), readline(), and readlines() methods to read content and the write() method to write content. For binary files, similar read() and write() methods can be used, but it is need to open the file in binary mode.

(5)File positioning. The seek(offset, whence) method is used to move the file pointer to a specified position. The offset represents the offset value, and whence specifies the starting position(0 for the beginning of the file, 1 for the current position, 2 for the end of the file). The tell() method returns the current position of the file pointer.

(6)Other file operations. The flush() method is used to immediately write the data from the buffer to the file without waiting for the buffer to be full or the file to be closed.

The truncate(size) method can adjust the size of the file. If the specified specified size is smaller than the cur-

当前大小小，则文件将被截断；如果指定的 size 比文件当前大小大，则文件将用空字节填充到指定大小。文件对象的 closed 属性可以检查文件是否已关闭。

（7）异常处理。在进行文件操作时，应该始终考虑异常处理，如 FileNotFoundError、IOError、PermissionError 等，以确保程序的健壮性。可以使用 try-except 块来捕获和处理这些异常，并提供适当的错误消息或恢复机制。

rent size, the file will be truncated; if the size is larger, the file will be padded with empty bytes to the specified size. The closed attribute of the file object can be used to check if the file is closed.

(7)Exception handling. When performing file operations, it is important to always consider exception handling, such as FileNotFoundError, IOError, PermissionError, to ensure the robustness of the program. One can use try-except blocks to catch and handle these exceptions, providing appropriate error messages or recovery mechanisms.

7.6 本章练习
7.6 Exercise for This Chapter

7.1 创建一个名为 example.txt 的文本文件，其中包含两行文本内容："The study of Python helps to enhance one's programming skills. Studying this course makes me happy."编写代码以读取该文件的所有内容，并将其打印出来。

7.2 搜索习题 7.1 中的 example.txt 文件是否包含字符串"Python"，并打印出包含该字符串的所有行。

7.3 文本文件合并与排序。假设有 3 个文本文件，每个文件包含一系列无序的整数，每行包含 1 个整数。将这些文件合并成一个新的文件，并将这些整数按升序排列，排序后每个整数各占一行。将排序后的整数写入一个新的文件 sorted_numbers.txt。

3 个文本文件中的内容如下：file1.txt 包含 3 个整数 9、3、5；file2.txt 包含 3 个整数 2、8、6；file3.txt 包含 3 个整数 1、7、4。

7.1 Create a text file named example.txt with two lines of text content: "The study of Python helps to enhance one's programming skills. Studying this course makes me happy.". Write code to read all the content of the file and print it out.

7.2 Search if the file example.txt from exercise 7.1 contains the string "Python", and print out all lines that contain this string.

7.3 Merging and sorting text files. Suppose there are three text files, each containing a series of unordered integers, with one integer per line. Merge these files into a new file and sort these integers in ascending order, with each integer on a separate line. Write the sorted integers to a new file named sorted_numbers.txt.

The contents of the three text files are as follows: file1.txt contains the three integers 9, 3, 5; file2.txt contains the three integers 2, 8, 6; file3.txt contains the three integers 1, 7, 4.

7.4 创建一个新文本文件 output.txt，并将一些文本内容写入该文件。例如，写入"Hello, world!"。

7.5 在习题 7.4 的 output.txt 文件末尾追加一段新的文本内容，例如"Another line of text."。

7.6 创建一个名为 data.txt 的文本文件，并写入 4 行文本，分别为"Line 1""Line 2""Line 3"和"Line 4"。读取 data.txt 文件中的第二行内容，并将其打印出来。

7.7 定位到习题 7.6 中 data.txt 文件的开始处，并覆盖写入新的第一行"New First Line"。

7.8 编写一个 Python 程序，将某个指定的字符串从文本文件中删除。例如，在 original_file.txt 文件中，有这样一段包含 5 行英文的英文文本："As the sun cast its golden rays over the vast expanse of green fields. A gentle breeze carried the scent of fresh earth and blooming flowers. In the distance. A group of sheep grazed peacefully. Their soft bleating sounds mingling with the chirping of birds and the rustling of leaves."删除其中的"sun"字符串，删除后的新内容保存在 new_file.txt 文件中。

7.4 Create a new text file named output.txt and write some text content into the file. For example, write "Hello, world!".

7.5 Append a new piece of text content, such as "Another line of text.", to the end of the file output.txt from exercise 7.4.

7.6 Create a text file named data.txt and write four lines of text, which are "Line 1""Line 2""Line 3" and "Line 4". Read the content of the second line from the data.txt file and print it out.

7.7 Position at the beginning of the file data.txt from exercise 7.6 and overwrite with a new first line "New First Line".

7.8 Write a Python program to remove a specified string from a text file. For example, in the file original_file.txt, there is a passage containing 5 lines of English text: "As the sun cast its golden rays over the vast expanse of green fields. A gentle breeze carried the scent of fresh earth and blooming flowers. In the distance. A group of sheep grazed peacefully. Their soft bleating sounds mingling with the chirping of birds and the rustling of leaves." Delete the string "sun" from the text and save the modified content in a new file named new_file.txt.

本章练习题参考答案

第 8 章 实战演练
Chapter 8 Practical Exercises

8.1 学生成绩管理系统
8.1 Student Performance Management System

本案例要求对 2020 级计算机专业学生进行成绩管理，假设该班级原始成绩数据有 10 条，后期可以对此班级的成绩信息进行扩充。学生成绩管理系统可以实现添加、查找、删除学生信息以及对其成绩进行管理等操作，并将保留在系统中的学生成绩信息保存在文本文件中以方便查阅。期望实现的功能是对文件中的数据进行增删改查操作，之后可对某一项成绩进行升序、降序排列，并将排序后的结果保存到新文件 stuSort.txt 中。如图 8-1 所示，是将 10 位学生按 Python 成绩升序排列之后的数据展示。

This case requires the management of grades for students majoring in Computer Science of the class of 2020. Assuming there are 10 original data entries for this class, the grade information for this class can be expanded later. The student grade management system should allow for adding, searching, deleting student information, and managing their grades. The student grade information retained in the system should be saved in a text file for easy reference. The expected functionality is to perform operations of adding, deleting, modifying, and searching data in the file. Subsequently, sorting a particular grade in ascending or descending order, and saving the sorted results to a new file named stuSort.txt. As shown in Figure 8-1, display the data after arranging the Python grades of 10 students in ascending order.

第8章 实战演练
Chapter 8　Practical Exercises

```
+---------------+---------------+----------------+---------------+-------------+
| Student ID    |     Name      | English Grades | Python Grades | Java Grades |
+---------------+---------------+----------------+---------------+-------------+
| 120100506010  |    Guo Jing   |       95       |      84       |     79      |
| 120100506008  | Liu Houqiang  |       76       |      85       |     69      |
| 120100506003  |    Yu Lili    |       93       |      86       |     79      |
| 120100506007  |   Yu Yanhui   |       84       |      86       |     92      |
| 120100506009  |    Lin Hui    |       95       |      86       |     74      |
| 120100506002  | Wang Jingjing |       92       |      91       |     89      |
| 120100506005  |  Gao Guoqing  |       93       |      91       |     92      |
| 120100506006  | Zhang Haonan  |       92       |      91       |     93      |
| 120100506001  |  Zhang Meili  |       88       |      92       |     95      |
| 120100506004  |  Bai Tingting |       95       |      92       |     91      |
+---------------+---------------+----------------+---------------+-------------+
```

图 8-1　按 Python 成绩升序排列的学生信息

Figure 8-1　Student Information Sorted in Ascending Order of Python Grades

运行此案例之后，本地文件夹中会出现一个文本文件，名为 student.txt，里面将会包含 10 位学生的个人信息以及 3 门专业课的成绩，如表 8-1 所示。

After running this case, a text file named student.txt will appear in the local folder. It will contain personal information of 10 students along with their grades for 3 major courses, as shown in Table 8-1.

表 8-1　学生成绩汇总

Table 8-1　Student Grades Summary

学号 Student ID	姓名 Name	英语成绩 English Grades	Python 成绩 Python Grades	Java 成绩 Java Grades
120100506001	Zhang Meili	88	92	95
120100506002	Wang Jingjing	92	91	89
120100506003	Yu Lili	93	86	79
120100506004	Bai Tingting	95	92	91
120100506005	Gao Guoqing	93	91	92
120100506006	Zhang Haonan	92	91	93
120100506007	Yu Yanhui	84	86	92
120100506008	Liu Houqiang	76	85	69
120100506009	Lin Hui	95	86	74
120100506010	Guo Jing	95	84	79

此案例将每个功能拆分成对应的函数，再在主函数中调用，方便读者参考学习。将增、删、改、查等操作都独立成函数，每一段代码实现一个具体的功能，保证程序的

In this case, each feature will be broken down into corresponding functions, which are then called in the main function for ease of reference and learning for the readers. Operations such as adding, deleting, modifying, and searching are each implemented as independent func-

可读性，其中程序的核心功能包括：

tions, with each segment of code accomplishing a specific function to ensure the readability of the program. The core functionalities of the program include:

```
def save(student_list)    #Save student grade information to a file
def insert()              #Enter student grade data
def search()              #Search for student data(support searching by ID or by name)
def delete()              #Delete data by student ID
def modify()              #Modify student grade data
def sort()                #Sort student grade data
```

8-1

接下来对实现学生成绩管理系统的函数逐一进行讲解。

Next, the writer will explain each function that implements the student grade management system one by one.

(1)使用 os 模块。

先对文件进行读取。os 是 Python 的内置模块，用于访问操作系统功能，在使用前需要使用 import os 语句导入。

import os.path 模块主要用于获取文件的属性，对于处理系统路径相关操作也是至关重要的。

(1)Using the "os" module.

First, read the file. The "os" module is a built-in module in Python that enables access to operating system functions. Before using it, one needs to import it using the "import os" statement.

The "import os.path" module is primarily used to retrieve file attributes and is also crucial for handling system path-related operations.

```
import os
import os.path
```

(2)定义一个指向保存学生基本信息以及成绩的文件对象。

(2)Define a file object pointing to the file where student basic information and grades are stored.

```
filename='student.txt'
```

(3)利用 save()函数保存学生成绩信息到文件中。

核心任务：定义一个 save()函数，接收一个列表类型的参数，列表包含学生的学号、姓名、3 门专业课成绩。

(3)Save student grade information to a file using the save() function.

Core task: Define a save function that accepts a list-type parameter. The list contains the student's ID, name, and scores for three major courses.

```
def save(student_list):
```

判断 student.txt 文件是否存在，如果不存在，则先在本地创建这个文件。

Check if the student.txt file exists, if it does not exist, create this file locally.

第8章　实战演练
Chapter 8　Practical Exercises

```
if not os.path.exists(filename):
    with open(filename,'w',encoding='utf-8') as file:
        pass
```

w：以写入方式打开一个文件。如果该文件已存在，则将其覆盖。如果该文件不存在，则创建新文件。

pass：占位符。此段代码的核心功能是检测本地是否存在名为 student.txt 的文件对象。

w: Open a file for writing. If the file already exists, it will be overwritten. If the file does not exist, a new file will be created.

pass: A placeholder. the core function of this code segment is to check if a file object named student.txt exists locally.

```
#Open the file and loop through the student grade information in the list to write it to the file
with open(filename, 'a', encoding='utf-8') as file:
    for stu in student_list:
        file.write(str(stu)+'\n')
```

a：以追加方式打开一个文件。如果该文件已存在，则文件指针将会放在文件结尾。也就是说，新的内容将会被写入已有内容之后。

注意：

①写入文件的内容必须是字符串类型，如果想存储数字（例如3门专业课的成绩、学号等），则需要先将其转成字符串类型；

②如果想写入的文件有换行，则需要添加换行符\n。

（4）使用 insert ()录入学生成绩。

创建一个空列表 student_list，用于存储学生成绩信息。首先进入循环，要求用户输入学生的学号、姓名、英语成绩、Python 成绩和 Java 成绩。如果用户没有输入学号或姓名，则跳过本次循环，继续等待用户输入。如果用户输入的成绩不是整数类型，则打印一条提示信息，继续等待用户输入。如果用户输入的信息无误，则将该学生的成绩以字典的形式保存到 student_list

a: Open a file for appending. If the file already exists, the file pointer will be positioned at the end of the file. In other words, new content will be written after existing content.

Note:

①The content written to the file must be of string type. If one wants to store numbers(such as grades for three major courses, student ID, etc.), one needs to convert them to string type first;

②If one wants the file to have line breaks, one needs to add the newline character \n.

(4)Enter student grades using the insert().

Create an empty list named student_list to store student grade information. Start a loop where the user is asked to input the student's ID, name, English score, Python score, and Java score. If the user does not input the ID or name, skip the current iteration and continue to wait for user input. If the user enters a score that is not an integer, print a message and continue to wait for user input. If the user input is correct, save the student's grades in dictionary format to the student_list. Ask the user if he wants to continue adding student information. If the user enters "y" or "Y", continue entering student

219

列表中。询问用户是否要继续添加学生信息。如果用户输入的是"y"或"Y"，则继续录入学生信息，否则跳出循环。将 student_list 列表作为参数传递给 save()函数，以便保存学生成绩信息。打印一条提示性语句，表示学生信息录入完毕。

information; otherwise, exit the loop. Pass the student_list as a parameter to the save() function to save the student grade information. Print a message indicating that the student information entry is complete.

```python
def insert():
    '''
    Enter student grade data
    return: None
    '''
    #List to store student grade information
    student_list=[]
    #Loop to input student grade data
    while True:
        id=input('Please enter the ID(e. g. 1001): ')
        if not id:
            continue
        name=input('Please enter the name: ')
        if not name:
            continue
        try:
            english=int(input('Please enter the English score: '))
            python=int(input('Please enter the Python score: '))
            java=int(input('Please enter the Java score: '))
        except:
            print('Invalid input, not an integer type. Please enter again')
            continue
        #Student grade data is stored in the form of a dictionary.
        studentDic={'id': id, 'name': name, 'english': english, 'python': python, 'java': java}
        #Add the student grade data to the list
        student_list.append(studentDic)
        answer=input('Do you want to continue adding? y/n')
        if len(answer)>0 and answer in 'yY':
            continue
        else:
            break
    save(student_list)
    print('Student information input completed')
```

第8章　实战演练
Chapter 8　Practical Exercises

（5）使用search()查看学生信息。

search()函数实现查找学生数据，并支持按照学号或姓名查找学生信息。用户通过输入数字选择按照学号查找还是按照姓名查找。

接下来打开存储学生信息的文件，文件名由变量filename保存，使用UTF-8编码读取文件内容，并将每一行存储到一个列表ls中。如果打开文件时发生异常，则打印一条错误提示信息。在用户选择按照学号查找时，程序会提示用户输入一个学号，然后使用列表推导式从列表ls中筛选出学号等于用户输入学号的学生信息，并将信息存储到一个列表t中。

如果找到了学生信息，则调用formatShow()函数显示查找结果；否则，打印一条信息提示没有找到学生信息。如果用户选择按照姓名查找，则实现方式同学号查找。用户选择除"1"或"2"的其他选项时，就退出循环，结束函数的执行。

(5) View student information using search().

The search() function is used to find student data and support searching student information by ID or name. The user selects whether to search by ID or name by inputting a number.

Next, the program opens a file containing student information, where the filename is stored in a variable called filename. The program reads the file content using UTF-8 encoding and stores each line in a list called ls. If an exception occurs while opening the file, an error message is printed. When the user chooses to search by ID, the program prompts the user to input an ID. It then uses a list comprehension to filter out the student information with ID matching the user-input ID and stores the information in a list called t.

If student information is found, the formatShow() function is called to display the search results; otherwise, a message indicating that no student information was found is printed. The same approach is followed when the user chooses to search by name. If the user selects an option other than "1" or "2", the loop is exited, and the execution of the function ends.

```python
def search():
    '''
    Search student data(support searching by ID or by name)
    return: None
    '''
    while True:
        select=input('To search by ID, enter 1. To search by name, enter 2. Enter any other key to exit:')
        try:
            #Open the student grades file and read all the data into a list named ls
            with open(filename, encoding='utf-8') as file:
                ls=file.readlines()
        except:
            print('An exception has occurred')
        #Search by ID
        if select=='1':
```

```
            id=input('Please enter the ID:')
            #Use list comprehension to find the data with the ID entered by the user.
            t=[d for d in [dict(eval(x)) for x in ls] if d['id']==id]
            if len(t)>0:
                formatShow(t)
            else:
                print(f'Student information with the ID {id} was not found')
        #Search by name
        elif select=='2':
            name=input('Please enter the name:')
            #Use list comprehension to find the data with the name entered by the user
            t=[d for d in [dict(eval(x)) for x in ls] if d['name']==name]
            if len(t)>0:
                formatShow(t)
            else:
                print(f'No student information found with the name {name}')
        else:
            break
```

（6）使用delete()按照学号删除数据。

delete()函数的主要功能是按照输入的学号删除相应的学生信息。代码执行流程如下：

进入while循环，在循环中通过input()函数输入要删除的学号，判断输入的学号是否为空字符串，如果不为空，则继续执行下一步。判断存储学生信息的文件是否存在，如果存在，则读取该文件中所有学生信息并将其存储在student_old列表中，否则student_old为空。

定义一个标志位flag，初始值为False，表示未找到要删除的学生信息，如果student_old不为空，则进入下一步；否则输出提示信息"No student information"并跳出while循环。

打开文件，遍历student_old列表中的所有数据，将其转换成字典类型，并将其存储到字典d中，如果d中的学号与输入的学号不相

(6) Delete data according to student ID using delete().

The main function of the delete() function is to delete the corresponding student information based on the input student ID. The code execution flow is as follows:

Entering the while loop, within the loop, use the input() function to enter the student ID to be deleted. Check if the input student ID is an empty string. If it is not empty, proceed to the next step. Check if the file storing student information exists. If the file exists, read all student information from the file and store it in a list called student_old, otherwise the student_old remains empty.

Define a flag variable flag with an initial value of False, indicating that the student information to be deleted has not been found. If student_old is not empty, proceed to the next step; otherwise, output a message "No student information" and break out of the while loop.

Open the file and iterate through all data in the student_old list, convert it into a dictionary, and store it in a dictionary called d. If the ID in d is not equal to the input ID, rewrite d to the file; otherwise, set flag to True,

Chapter 8　Practical Exercises

等，则将 d 重新写入文件中；否则将 flag 置为 True，表示找到了要删除的学生信息。判断 flag 的值，如果为 True，则输出删除成功的提示信息；否则输出没有找到要删除的学生信息的提示信息。

执行 show() 函数，显示学生信息。通过 input() 函数输入是否继续删除的答案，如果是，则继续删除；否则跳出 while 循环。

indicating that the student information to be deleted has been found. Check the value of flag; if it is True, output a message indicating successful deletion; otherwise, output a message indicating that the student information to be deleted was not found.

Execute the show() function to display student information. Use the input() function to input the answer on whether to continue with the deletion. If yes, continue with the deletion; otherwise, break out of the while loop.

```
def delete():
    '''
    Delete data based on the student ID
    return: None
    '''
    while True:
        id=input('Please enter the ID of the student you want to delete: ')
        if id != '':
            #Open the student performance file and read all the data into the list student_old
            if os.path.exists(filename):
                with open(filename, 'r', encoding='utf-8') as file:
                    student_old=file.readlines()
            else:
                student_old=[]
            flag=False
            #Flag variable, assuming False represents that the data to be deleted was not found
            #If student_old is not empty
            if student_old:
                with open(filename, 'w', encoding='utf-8') as file:
                    d={}
                    for item in student_old:
                        d=dict(eval(item))
                        #Rewrite the IDs that are not the one to be deleted back into the file
                        if d['id'] != id:
                            file.write(str(d)+'\n')
                        #Indicate that the data to be deleted was found in the list by setting the flag to True
                        else:
                            flag=True
                if flag:
                    print(f'The student information with ID {id} has been deleted')
                else:
```

```
                    print(f'The student information with ID {id} was not found')
        else:
    #If student_old is empty, it means there is no student performance data in the file
            print('No student information')
            break
        #Display student grade data
        show()
        answer=input('Do you want to continue deleting? y/n')
        if answer=='y' or answer=='Y':
            continue
        else:
            break
```

(7) 使用 modify()修改学生成绩数据。

首先，通过 show()函数显示所有学生成绩数据。然后，要求用户输入要修改学生的学号。如果输入的学号不为空，则会打开学生成绩文件并读取其中的数据。在读取的过程中，将每个学生成绩数据转换成字典，并将这些字典保存在一个 ls 列表中。

接下来，通过列表推导式找到要修改的数据，并将其输出到屏幕上。

然后，要求用户输入新的姓名、英语成绩、Python 成绩和 Java 成绩，并将其保存到要修改的学生数据中。

最后，将修改后的所有数据都写回到学生成绩文件中，并询问用户是否继续修改其他学生的数据。

如果用户输入的学号为空，则提示用户学生学号不能为空。如果学生成绩文件不存在，则提示无法进行修改操作。如果要修改的学生学号不存在，则输出提示信息并要求用户重新输入要修改学生的学号。

(7)Modify student grade data with modify().

First, display all student grade data using the show() function. Then, prompt the user to input the ID of the student to be modified. If the entered ID is not empty, open the student grade file and read the data from it. During the reading process, convert each student grade data into a dictionary and store these dictionaries in a list named ls.

Next, use list comprehension to find the data to be modified and output it on the screen.

Then, prompt the user to input the new name, English grade, Python grade, and Java grade, and save them to the student data to be modified.

Finally, write back all the modified data to the student grade file and ask the user if he wants to continue modifying data for other students.

If the ID entered by the user is empty, prompt the user that the student ID cannot be empty. If the student grade file does not exist, inform that the modification operation cannot be performed. If the student ID to be modified does not exist, output a prompt message and ask the user to re-enter the ID to be modified.

第8章 实战演练
Chapter 8　Practical Exercises

下面这段代码使用了循环语句和条件语句来实现功能，同时也用到了文件读写和字典等数据类型的操作。

This code snippet utilizes loops and conditional statements to achieve functionality, while also involving file reading and writing as well as operations with data types such as dictionaries.

```python
def modify():
    '''
    Modify student grade data
    return: None
    '''
    while True:
        #Display student grade data
        show()
        id=input('Please enter the ID of the student you want to modify:')
        if id!='':
            if os.path.exists(filename):
                ls=[]
                #Open the student grade file and read all data into the liststudent_old
                with open(filename, 'r', encoding='utf-8') as file:
                    student_old=file.readlines()
                    #Iterate through student_old, convert each student grade data into a dictionary, and
                    #save them in the list ls
                    for item in student_old:
                        d=dict(eval(item))
                        ls.append(d)
                #Using list comprehension to find the data to be modified
                t=[x for x in ls if x['id']==id]
                if len(t)>0:
                    print('You have found the student, and now you can modify his information')
                    formatShow(t)
                    name=input('Please enter the name: ')
                    if not name:
                        break
                    english=input('Please enter the English score: ')
                    python=input('Please enter the Python score: ')
                    java=input('Please enter the Java score: ')
                    #Only modify the first matching data, ignore the matching data from
                    #the second one onward
                    t[0]['name']=name
                    t[0]['english']=english
                    t[0]['java']=java
                    #Write the modified data as well as other data back to the file
                    with open(filename, 'w', encoding='utf-8') as file:
```

225

```
                    for stu in ls:
                        file.write(str(stu)+' \ n')
                    print('Modification completed')
                    answer=input('Do you want to continue with the modifications? y/n \ n')
                    if answer=='y' or answer=='Y':
                        continue
                    else:
                        break
            else:
                print(f'There is no student information with the ID {id}')
        else:
            print('The file does not exist, unable to proceed with the modification operation')
    else:
        print('The student ID cannot be empty')
```

（8）使用 sort() 对学生成绩进行升序或者降序排列。

定义一个名为 sort() 的函数，其主要功能是对学生成绩数据进行排序，并将排序结果保存到新文件中。

首先，该函数会检查名为 filename 的文件是否存在，如果不存在则会打印提示信息并返回。如果文件存在，则会创建一个空列表 ls，并使用 with 语句打开文件，读取每行数据，将其转换成字典格式后添加到 ls 列表中。

接下来，该函数调用 show() 函数展示所有学生的成绩信息，并提示用户选择升序或降序排序以及排序方式。根据用户的输入，该函数会使用列表内置排序方法 sort() 对数据进行排序。具体来说，如果用户选择按英语成绩排序，则会以 english 键对应的值为排序关键字进行排序；如果选择按总成绩排序，则会以所有 3 门课程成绩之和为排序关键字进行排序。

排序完成后，该函数会调用

(8) To sort the students' grades in ascending or descending order, one can use the sort() function.

This code defines a function called sort(), which is mainly used to sort the students' grade data and save the sorted results to a new file.

Firstly, the function checks if the file with the name "filename" exists. If it does not exist, it prints a prompt message and returns. If the file exists, it creates an empty list called ls, and then uses the "with" statement to open the file, read each line of data, convert it into dictionary format, and add it to the ls list.

Next, the function calls the show ( ) function to display all students' grade information and prompts the user to choose either ascending or descending order for sorting, as well as the sorting method. Based on the user's input, the function sorts the data using the list's built-in sort() method. Specifically, if the user chooses to sort by English scores, the sorting key will be the value associated with the "english" key; if the user selects to sort by total scores, the sum of the scores of all three courses will be used as the sorting key.

After sorting is completed, the function calls the for-

formatShow()函数展示排序结果，并调用 writefile()函数将排序结果保存到名为 stuSort.txt 的新文件中。

最后，该函数会询问用户是否继续操作，如果用户选择继续，则会再次进行排序操作；否则函数执行结束。

总的来说，该函数的执行流程为：检查文件是否存在→读取文件中的数据并将其转换为字典格式→展示所有学生的成绩信息→提示用户选择排序方式→对数据进行排序→展示排序结果→保存排序结果到新文件中→询问用户是否继续操作→如果选择继续则重复上述过程；否则函数执行结束。

按总成绩排序的关键语句是：

matShow() function to display the sorted results and saves the sorting results to a new file named stuSort.txt, achieved by calling the writefile() function.

Lastly, the function will ask the user if he wishes to continue the operation. If the user chooses to continue, another sorting operation will be carried out; otherwise, the function execution will end.

In summary, the execution flow of this function is as follows: check if the file exists→read the data from the file and convert it into dictionary format→display all students' grade information→prompt the user to choose the sorting method→sort the data→display the sorting results→save the sorting results to a new file→ask the user if he wishes to continue the operation→if he chooses to continue, repeat the above process; otherwise, end the function execution.

The key statement for sorting by total scores is:

ls.sort(key=lambda x:(int(x['english'])+int(x['python'])+int(x['java'])), reverse=flag)

中括号[]内为成绩的标识，各科总分数转整数进行升序或者降序操作：0 表示升序，1 表示降序。将 10 名同学按 Java 成绩升序排列的统计结果如图 8-2 所示。

In the square brackets [ ], the identifiers for the grades are specified. The total scores for each subject are converted to integers for ascending or descending operations: Click 0 for ascending and 1 for descending. the statistical results of sorting the Java scores of 10 students in ascending order are shown in Figure 8-2.

```
+---------------+---------------+---------------+---------------+---------------+
| Student ID    |     Name      | English grades| Python grades | Java grades   |
+---------------+---------------+---------------+---------------+---------------+
| 120100506008  | Liu Houqiang  |      76       |      85       |      69       |
| 120100506009  |   Lin Hui     |      95       |      86       |      74       |
| 120100506003  |   Yu Lili     |      93       |      86       |      79       |
| 120100506010  |   Guo Jing    |      95       |      84       |      79       |
| 120100506002  | Wang Jingjing |      92       |      91       |      89       |
| 120100506004  | Bai Tingting  |      95       |      92       |      91       |
| 120100506005  | Gao Guoqing   |      93       |      91       |      92       |
| 120100506007  |  Yu Yanhui    |      84       |      86       |      92       |
| 120100506006  | Zhang Haonan  |      92       |      91       |      93       |
| 120100506001  | Zhang Meili   |      88       |      92       |      95       |
+---------------+---------------+---------------+---------------+---------------+
```

图 8-2 将 10 名同学按 Java 成绩升序排列的统计结果

Figure 8-2 The Statistical Results of Sorting the Java Scores of 10 Students in Ascending Order

```
def sort():
    '''
    Sort the student performance data
```

```python
        return: None
        '''
        if not os.path.exists(filename):
            print('The file does not exist')
            return
        ls=[]
        with open(filename, 'r', encoding='utf-8') as file:
            t=file.readlines()
            if not t:
                print('No data available to perform the query')
                return
            else:
                for item in t:
                    ls. append(dict(eval(item)))
        show()
        while True:
            mode=input('Please select(0 for ascending, 1 for descending)')
            if mode=='0':
                flag=False
            elif mode=='1':
                flag=True
            else:
                print('Invalid input, please try again')
                continue
            sub=input('Please select the sorting method(1. Sort by English score, 2. Sort by Python score, 3. Sort by Java score, 0. Sort by total score)')
            #According to the user input, use the built-in list sorting method sort() to sort the data
            if sub=='1':
                ls. sort(key=lambda x: int(x['english']), reverse=flag)
            elif sub=='2':
                ls. sort(key=lambda x: int(x['python']), reverse=flag)
            elif sub=='3':
                ls. sort(key=lambda x: int(x['java']), reverse=flag)
            elif sub=='0':
                ls. sort(key=lambda x:(int(x['english'])+int(x['python'])+int(x['java'])), reverse=flag)
            else:
                print('Invalid input, please enter again')
                continue
            formatShow(ls)
    #Create a new file object "stuSort.txt", save the sorted results to the new file, and call the
    #writefile() function to achieve this
            new_file='stuSort.txt'
            writefile(ls, new_file)
```

```
answer=input('Do you want to continue?(y/n)')
if answer=='y' or answer=='Y':
    continue
else:
    break
```

（9）使用 showMenu()展示菜单管理界面。

showMenu()函数的作用是显示一个学生成绩管理系统的菜单，其包括以下功能。

①录入学生信息。
②查找学生信息。
③删除学生信息。
④修改学生信息。
⑤排序。
⑥统计学生总人数。
⑦显示所有学生信息。
⑧退出。

在函数执行时，会先打印一段学生成绩管理系统的标题，然后再打印出上述菜单选项。该函数没有返回值，直接结束。

(9) Display the menu management interface using showMenu().

The showMenu() function is used to display a menu for a student grade management system, including the following functions.

①Enter student information.
②Search for student information.
③Delete student information.
④Modify student information.
⑤Sort.
⑥Calculate the total number of students.
⑦Display all student information.
⑧Exit.

When the function is executed, it will first print a title for the student grade management system, then print the menu options mentioned above. The function does not return any value, it simply ends.

```
def showMenu():
    '''
    Display the student grade management system menu
    return: None
    '''
    print('===============Student Grade Management System===============')
    print('------------------------- Function Menu -------------------------')
    print('1. Enter student information')
    print('2. Search for student information')
    print('3. Delete student information')
    print('4. Modify student information')
    print('5. Sorting')
    print('6. Calculate the total number of students')
    print('7. Display all student information')
    print('0. Exit')
    print('----------------------------------------------------------------')
```

(10) 主函数 main() 构建。

定义一个 main() 函数,用于运行整个程序。在 main() 函数中,使用一个无限循环"while True:",保证程序一直运行,直到用户选择退出系统。

在循环中调用 showMenu() 函数,显示菜单选项,让用户选择需要执行的操作。通过 input() 函数获取用户输入的选项,并将其转换为整数类型赋值给 choice 变量。

判断用户输入是否在范围内(0~7),如果不在,则提示用户重新输入。如果用户输入 0,则表示用户想要退出系统,此时程序会弹出一个询问框,让用户确认是否真要退出系统。如果用户输入的选项为 1~7,则表示用户选择了对应的操作。根据用户输入的选项,调用相应的函数执行对应的操作。如果用户输入不是 0~7 的数字,则程序会提示用户重新输入。主菜单如图 8-3 所示。

(10)Construct the main function called main().

Define a main() function to run the entire program. Use an infinite loop "while True:" in the main() function to ensure that the program keeps running until the user chooses to exit the system.

Within the loop, call the showMenu() function to display the menu options for the user to choose the operation he wants to perform. Use the input() function to obtain the user's choice, convert it to an integer, and assign it to the "choice" variable.

Check if the user's input is within the range of 0 to 7. If it is not within this range, prompt the user to re-enter the input. If the user inputs 0, it indicates that the user wants to exit the system. At this point, the program will display a dialogue box asking the user to confirm if he really wants to exit the system. If the user inputs 1 to 7, it means the user has selected the corresponding operation. Depending on the user's input, call the relevant function to perform the corresponding operation. If the user's input does not fall within the range of 0 to 7, the program will prompt the user to re-enter the input. The Main Menu is shown in Figure 8-3.

```
========================Student Grade Management System========================
--------------------------Function Menu--------------------------
1. Enter student information
2. Search for student information
3. Delete student information
4. Modify student information
5. Sorting
6. Calculate the total number of students
7. Display all student information
0. Exit
-----------------------------------------------------------------
Please select:
```

图 8-3 主菜单

Figure 8-3 Main Menu

其中,不同选项对应的操作如下:

The operations corresponding to each option are as follows:

```
def main():
    while True:
```

```
            showMenu()
            choice=int(input('Please select:'))
            if choice in range(8):
                if choice==0:
                    answer=input('Are you sure you want to exit the system? y/n:')
                    if len(answer)>0 and answer in 'Yy':
                        print('Thank you for using the system!')
                        break
                elif choice==1:
                    insert()
                elif choice==2:
                    search()
                elif choice==3:
                    delete()
                elif choice==4:
                    modify()
                elif choice==5:
                    sort()
                elif choice==6:
                    totalNumStu()
                elif choice==7:
                    show()
            else:
                print('Please enter a number menu between 0 and 7.')
if __name__=='__main__':
    main()
```

8.2 学生学业成绩统计
8.2 Student Academic Performance Statistics

这一小节里，我们将学习使用 pandas 库来读 Excel 文件，使用 matplotlib 库来绘制饼图和条形图。本案例要求对期末考试成绩进行统计，并绘制出相应的饼图和条形图。

需要安装的第三方库有 pandas 和 matplotlib，由于使用 pandas 库读取 Excel 文件需要依赖 xlrd 库，所以本案例还需要安装 xlrd 库。安装命令如下：

In this section, we will learn how to use the pandas library to read Excel files and use the matplotlib library to plot pie charts and bar graphs. This case study requires statistical analysis of final exam grades and the creation of corresponding pie charts and bar graphs.

The third-party libraries that need to be installed are pandas, matplotlib, and xlrd (required for reading Excel files with pandas). The installation command is as follows:

```
pip install pandas
pip install matplotlib
pip install xlrd
```

首先，设计存储学生成绩数据的 Excel 文件，Excel 文件内容格式如图 8-4 所示，包含姓名和成绩两列数据。

Firstly, design the student performance data using an Excel file with the format as shown in Figure 8-4, including two columns of data: Name and Grades.

	A	B
1	Name	Grades
2	Zhang Meili	75
3	Wang Jingjing	92
4	Yu Lili	93
5	Bai Tingting	95
6	Gao Guoqing	88
7	Zhang Haonan	92
8	Yu Yanhui	84
9	Liu Houqiang	76
10	Lin Hui	66
11	Guo Jing	55

图 8-4　Excel 文件内容格式

Figure 8-4　Excel File Content Format

运行结果包含以下 3 方面内容。

（1）将统计结果写到文本文件中。

（2）根据统计结果绘制饼图和条形图。

（3）统计结果文件示例如下所示，包括总人数、最高分、最低分、平均分等数据。

The run results include three aspects.

(1)Write the statistical results to a text file.

(2)Plot a pie chart and a bar graph based on the statistical results.

(3)The statistical results file is as follows, including data such as total number of students, highest score, lowest score, average score, etc.

```
成绩统计结果：
总人数：10
最高分：95.0
最低分：55.0
平均分：81.6
优秀：4 人(40.00%)
良好：2 人(20.00%)
中等：2 人(20.00%)
及格：1 人(10.00%)
不及格：1 人(10.00%)
```

```
Grade statistics result:
Total:10
Highest score:95.0
Lowest score:55.0
Average score:81.6
Excellent:4 people(40.00%)
Good:2 people(20.00%)
Average:2 people(20.00%)
Pass:1 people(10.00%)
Fail:1 people(10.00%)
```

第8章　实战演练
Chapter 8　Practical Exercises

绘制的饼图和条形图如图 8-5 和图 8-6 所示。

The pie chart and bar graph are shown in Figure 8-5 and Figure 8-6, respectively.

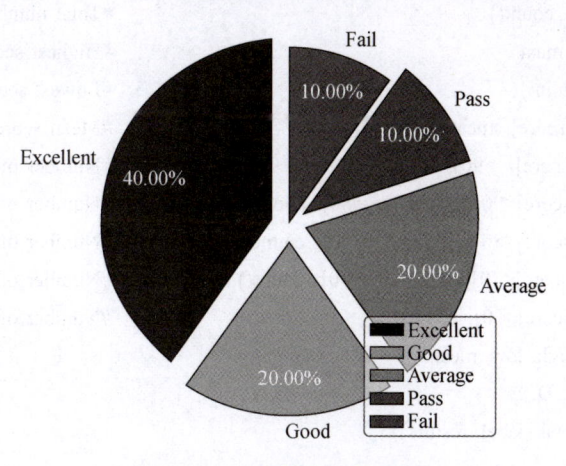

图 8-5　成绩分布饼图

Figure 8-5　Grade Distribution Pie Chart

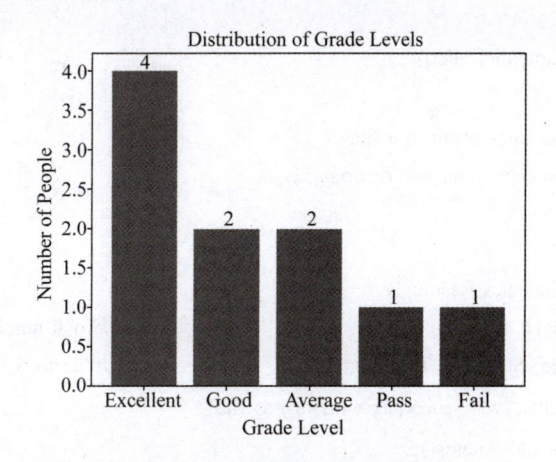

图 8-6　成绩分布条形图

Figure 8-6　Grade Distribution Bar Graph

完整参考代码如下：

Below is the complete reference code:

```
import os
import pandas as pd
import matplotlib.pyplot as plt


def cal(file_path: str):
    """
    Read Excel file, analyze grade data
    param file_path: Excel file path
    return: None
    """
```

233

```python
        df=pd.read_excel(file_path)
        score='Grades'
        count=df[score].count()                              #Total number of people
        high=df[score].max()                                 #Highest score
        low=df[score].min()                                  #Lowest score
        mean=round(df[score].mean(), 2)                      #Mean score
        A=df[score][df[score]>=90].count()                   #Number of A grades
        B=df[score][df[score]<90][df[score]>=80].count()     #Number of B grades
        C=df[score][df[score]<80][df[score]>=70].count()     #Number of C grades
        D=df[score][df[score]<70][df[score]>=60].count()     #Number of D grades
        E=df[score][df[score]<60].count()                    #Number of E grades
        result1=count, high, low, mean
        result2=A, B, C, D, E
        dict1=dict(zip(total_label, result1))
        dict2=dict(zip(score_label, result2))
        return dict1, dict2

def write_file(result: tuple[dict, dict]):
    """
    Write the grade statistics result to a file
    param result: tuple containing two dictionaries
    return: None
    """
    print('\nGrade Statistics Result: ')
    count=sum(result[1].values())                            #Total number of students
    file_name='Grade_Statistics_Result.txt'                  #File to write the statistics result
    with open(file_name, 'w+', encoding='utf-8') as file:
        for k, v in result[0].items():
            file.write(f'{k}: {v}\n')
        for k, v in result[1].items():
            file.write(f'{k}: {v} people({v / count*100:.2f}%)\n')
        file.seek(0)
        print(file.read())
    print(f'Grade statistics result has been written to the file: {os.path.join(os.getcwd(), file_name)}')

def show_pie(result: dict):
    """
    Display a pie chart
    param result: dictionary containing data for the pie chart
    return: None
    """
    #Display Chinese characters correctly
```

```python
        plt.rcParams['font.family']='SimHei'
        explode=0.1,0.1,0.1,0.1,0              #Explode values for each part
        plt.pie(list(result.values()),explode=explode,labels=list(result.keys()),labeldistance=1.1,
autopct='%.2f%%',shadow=False,startangle=90,pctdistance=0.8)
        plt.title('Distribution of Grade Levels',fontproperties='SimHei',fontsize=20)
        plt.legend(loc='lower right')
        plt.show()

    def show_bar(result: dict):
        """
        Display a bar chart
        param result: dictionary containing data for the bar chart
        return: None
        """
        #Display Chinese characters correctly
        plt.rcParams['font.family']='SimHei'
        rects=plt.bar(list(result.keys()),list(result.values()))
        autolabel(rects=rects)
        plt.title('Distribution of Grade Levels',fontproperties='SimHei',fontsize=20)
        plt.xlabel('Grade Level',fontproperties='SimHei',fontsize=20)
        plt.ylabel('Number of People',fontproperties='SimHei',fontsize=20)
        plt.show()

    def autolabel(rects):
        """
        Automatically label the y-axis values of the bar chart
        param rects: List of Rectangle class objects to be labeled
        return: None
        """
        for rect in rects:
            height=rect.get_height()
            plt.annotate(height,
                xy=(rect.get_x()+rect.get_width()/2, height),
                xytext=(0, 2),
                textcoords='offset points')

    if __name__=='__main__':
        total_label=['Total', 'Highest Score', 'Lowest Score', 'Average Score']
        score_label=['Excellent', 'Good', 'Average', 'Pass', 'Fail']
        print('Excel file format requirement for scores: ')
        print('Name', 'Score')
        print('John Doe', 90)
```

```
        print('Jane Smith',85)
        print('…', '…')
    tip=r'Please enter the path of the Excel file with scores(e. g. , D:\FinalExamScores.xlsx):'
    file=input(tip)
    while not file:
        file=input(tip)
    try:
        cal_result=cal(file)      #Call the calculation function
        write_file(cal_result)
        input('Press Enter to display the pie chart and bar chart')
        show_pie(cal_result[1])
        show_bar(cal_result[1])
    except Exception as e:
        print('An exception occurred:', e)
    input()    #Wait for user input to avoid the program from exiting directly after running
```

代码分析。

(1) 导入标准库和第三方库，代码如下：

Code analysis.

(1) Import the standard and third-party libraries with the following code:

```
import os
import pandas as pd
import matplotlib.pyplot as plt
```

(2) 定义 cal()函数，使用 pandas 库中的 read_excel()方法读取 Excel 文件，代码为"df = pd. read_excel(filePath)"，df 数据类型为 DataFrame 类型，统计总人数可以使用代码"count = df['Grades'].count()"。统计最高分使用 max()方法，"df['Grades'].max()"对 df 中成绩列获取最大值。同理可以使用 min()方法获取最小值、mean()方法计算平均分。统计大于或等于 90 分的学生人数可以使用条件筛选：

(2) Define the cal() function. Use the read_excel() method from the pandas library to read the Excel file with the code "df=pd.read_excel(filePath)", The "df" variable holds data in the form of a DataFrame. To calculate the total number of people, one can use the following code "count=df['Grades']. count()". To calculate the highest score, use the max() method: "df['Grade']. max()". This code retrieves the maximum value in the "Grades" column. Similarly, one can use the min() method to get the minimum value and the mean() method to calculate the average score. To count the number of students with scores greater than or equal to 90, use:

```
df['Grades'][df['Grades']>=90].count()
```

统计分数在 80 到 90 之间学生人数，可以使用条件叠加操作：

To count the number of students with scores between 80 and 90, one can use a combination of conditions:

df['Grades'][df['Grades']<90][df['Grades']>=80].count()

cal()函数会计算统计结果，并将多个统计结果以元组形式返回。

（3）write_file()函数接收 cal()函数统计结果，先将统计结果在控制台输出，再将统计结果写入文件。

（4）show_pie()函数接收 cal()函数统计结果的第二个值（字典类型），包含优秀、良好、中等、及格、不及格每个等级的人数，绘制成饼图。

（5）show_bar()函数和 show_pie()函数功能类似，接收 cal()函数统计结果的第二个值（字典类型），包含优秀、良好、中等、及格和不及格每个等级的人数，绘制成条形图。

（6）autolabel()函数对条形图 y 轴值进行标注。

The cal() function computes the statistical results and returns multiple statistics as a tuple.

(3) The write_file() function receives the statistical results from the cal() function, outputs the statistics to the console, and then writes the results to a file.

(4) The show_pie() function receives the second value(dictionary type) of the statistical result from the cal() function, which includes the number of people at each level of excellent, good, average, pass, and fail, and plots it as a pie chart.

(5) The functions show_bar() and show_pie() have similar functions. It receive the second value(dictionary type) of the statistical results from the cal() function, which includes the number of people at each level, including excellent, good, average, pass, and fail, and plots it as a bar chart.

(6) The autolabel() function annotates the y-axis values of the bar graph.

8.3　绘制词云
8.3　Generating Word Cloud

这一小节里，我们将学习使用第三方库 wordcloud 来绘制词云。本案例将使用"中国共产党第二十次全国代表大会报告"中文版和英文版分别绘制中文词云图和英文词云图。

需要安装的第三方库有 wordcloud、matplotlib 和 jieba 库。安装命令如下：

In this section, we will learn how to create a word cloud using the third-party library "wordcloud". This case will use the Chinese and English versions of "Report to the 20th National Congress of the Communist Party of China" to create Chinese and English word cloud images respectively.

The third-party libraries needed to be installed include wordcloud, matplotlib, and jieba. The installation commands are as follows:

```
pip install wordcloud
pip install matplotlib
pip install jieba
```

首先准备好中文版和英文版的"中国共产党第二十次全国代表大会报告"原文，分别保存在文本文件"二十大报告.txt"和"Report to the 20th Congress.txt"中。程序分别读取这两个文本文件，根据文件内容绘制中文词云图和英文词云图，运行结果如图8-7和图8-8所示。

First, prepare the original Chinese and English versions of "Report to the 20th National Congress of the Communist Party of China" and save them in text files named "二十大报告.txt" and "Report to the 20th Congress.txt", respectively. The program will then read these two text files, generate Chinese and English word clouds based on the file content, and the running result will be displayed as shown in Figures 8-7 and Figure 8-8.

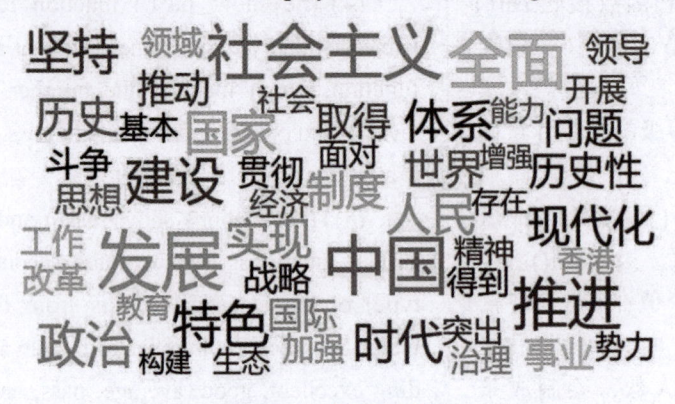

图 8-7 中文词云图

Figure 8-7 Chinese Word Cloud

图 8-8 英文词云图

Figure 8-8 English Word Cloud

完整参考代码如下：

The complete reference code is as follows:

```
import jieba.analyse
import matplotlib.pyplot as plt
from wordcloud import WordCloud

def readFile(fileName):
    """
    Read the text file one wants to analyze
```

```
    param fileName: File name, including the full path information
    return: File content
    """
    with open(fileName, 'r', encoding='gbk') as f:
        text=f.read()
    return text

def analysisWords(text):
    """
    Analyze strings, count the frequency of each word appearing in the text
    param text: String
    return: Word frequency count results, returned in dictionary form
    """
    #The word segmentation method uses jieba. analyse. textrank()
    result=jieba. analyse. textrank(text, topK=50, withWeight=True)
    wordDic={}
    t=0
    for i in result:
        wordDic[i[0]]=i[1]
        t +=i[1]
    return wordDic

def draw(word, isChinese):
    """
    Draw a word cloud
    param word: English string or dictionary(key: keyword, value: frequency or weight)
    param isChinese: Is it Chinese
    return: None
    """
    wordCloud=WordCloud(font_path='msyh. ttc',    #Font style path
            background_color=None,        #Set the background color of the word cloud
            max_font_size=200,            #Set the maximum font size in the word cloud
            scale=1. 5,                   #Scale up the canvas proportionally
            max_words=50,                 #Maximum number of words to display
            width=1980,                   #Output image width
            height=1080,                  #Output image height
            mode='RGBA',                  #Transparent background will be
#generated when mode is RGBA and background_color is None
            prefer_horizontal=1,          #The displayed text is not rotated
            repeat=False)                 #Display number of words
    if isChinese:
        wordCloud. generate_from_frequencies(word)
```

```
        else:
            wordCloud.generate(word)
        #Display word clouds using matplotlib
        plt.imshow(wordCloud, interpolation='bilinear')
        plt.axis('off')   #Do not display axes
        plt.show()

if __name__=='__main__':
    #To generate a word cloud, first prepare the text file containing the content one wants to
    #analyze. One can choose to use a different file for analysis as needed, and then
    #proceed to create the word cloud image
        #Create an English word cloud
        #Create a Chinese word cloud
        fileName='二十大报告.txt'
        txt=readFile(fileName)
        wordDic=analysisWords(txt)
        draw(wordDic, True)
        fileName='Report to the 20th Congress.txt'
        txt=readFile(fileName)
        draw(txt, False)
```

代码分析。

（1）导入第三方库，代码如下：

Code analysis.

(1)Import the third-party library, the code is as follows:

```
import jieba.analyse
import matplotlib.pyplot as plt
from wordcloud import WordCloud
```

（2）readFile()函数的功能是读取文本文件内容，读取编码格式为 GBK 格式。要求文本文件编码格式为 ANSI 格式，否则读取出来的内容可能会有乱码。

（3）analysisWords()函数的功能是对中文进行处理，使用 jieba 库中 textrank()方法提取中文关键词和对应的权重值。函数返回值是字典类型，为文本中前 50 个关键词以及对应的权重值。

（4）draw()函数的功能是绘制词云，如果文本内容是中文，需要

(2)The function readFile() is used to read the content of a text file with the encoding format set as GBK. The text file must be encoded in ANSI format, otherwise, the content read may appear as garbled text.

(3)The function analysisWords() processes Chinese text by using the textrank() method from the "jieba" library to extract Chinese keywords and their corresponding weights. The function returns a dictionary containing the top 50 keywords from the text along with their weights.

(4)The function draw() is responsible for drawing the word cloud. If the text content is in Chinese, a

绘制中文词云，使用 wordcloud 库中的 generate_from_frequencies()方法绘制，接收的参数为步骤(3)中生成的字典类型数据。如果文本内容是英文，需要绘制英文词云，使用 wordcloud 库中 generate()方法即可，接收参数为英文原文。

Chinese word cloud needs to be generated using the generate_from_frequencies() method from the "wordcloud" library, with the input parameter being the dictionary data generated in step(3). If the text content is in English, an English word cloud can be generated using the generate() method from the "wordcloud" library with the English original text as the input.

8.4 绘制七段管
8.4 Drawing a Seven-segment Display

这一小节里，我们将学习使用第三方库 turtle 绘制七段管，来显示当前日期时间。七段管，借由 7 个发光二极管以不同组合来显示数字，又称为七划管或七段数码管。由于所有灯管全亮时所表示的是 8，所以又称 8 字管。使用 turtle 库绘制的七段管如图 8-9 所示。

In this section, we will learn to use the third-party "turtle" library to draw a seven-segment display to show the current date and time. The seven-segment display, also known as the "seven-bar display" or "seven-segment digital display", displays numbers by different combinations of seven light-emitting diodes. It is also called the "8-segment display" because when all the segments are lit up, it represents the number "8". The seven-segment display drawn using the "turtle" library is shown in Figure 8-9.

图 8-9 turtle 绘制的七段管

Figure 8-9 Seven-segment Display Drawn Using Turtle Library

每个数字由 7 条线段组成，线段可能绘制，也可能不绘制，绘制结束后所表示的数字不同。例如，如果 7 条线段都进行绘制，则表示的是数字 8；如果中间位置的线段不绘制，则表示的是数字 0，以此类推。如图 8-10 所示，对 7 条线段进行编号。在绘制时，从 1 开始绘制，一直绘制到 7，当前数字即绘制完毕。绘制过程为：绘制 1（如果需要则移动绘制，否则只移动不绘制），右转 90 度，绘制 2

Each digit is composed of 7 segments, which may be drawn or left undrawn, resulting in different displayed numbers after drawing. For example, if all 7 segments are drawn, it represents the number 8; if the middle segment is not drawn, it displays the number 0, and so on. Figure 8-10 shows the numbering of the 7 segments. When drawing, start from 1 and continue drawing up to 7, and the current digit is completed. The drawing process is as follows: draw 1(move and draw if needed, otherwise only move), turn right 90 degrees, draw 2(move and draw if needed, otherwise only move), turn right 90 degrees, draw 3, and so on.

（如果需要则移动绘制，否则只移动不绘制），右转90度，绘制3……

图8-10　七段管的7条线段

Figure 8-10　7 Segments of the Seven-segment Display

本案例使用turtle库绘制，所用到的方法介绍如下。

（1）turtle.penup()：抬起画笔。

（2）turtle.fd(distance)：沿当前方向移动distance距离。

（3）turtle.pendown()：落下画笔。

（4）turtle.right(angle)：顺时针旋转angle角度。

（5）turtle.left(angle)：逆时针旋转angle角度。

（6）turtle.pencolor(*args)：设置画笔颜色。

（7）turtle.write(arg)：将arg对象写到屏幕上，本例中用来写"年""月""日"和"："。

（8）turtle.setup()：设置主窗口大小和位置。

（9）turtle.speed(speed=None)：设置海龟绘制时的移动速度。speed可以取值0～10，从1到10速度逐渐增加，0比较特殊，代表速度最快。

（10）turtle.pensize(width=None)：设置画笔线条的粗细。

（11）turtle.done()：绘制结束，维持界面不退出，调用Tkinter的mainloop()函数。

This example utilizes the turtle library for drawing, and the methods used are described as follows.

(1)turtle.penup(): Lift the pen up.

(2)turtle.fd(distance): Move the turtle forward by the specified distance.

(3)turtle.pendown(): Put the pen down to continue drawing.

(4)turtle.right(angle): Rotate the turtle clockwise by the specified angle.

(5)turtle.left(angle): Rotate the turtle counterclockwise by the specified angle.

(6)turtle.pencolor(*args): Set the pen color.

(7)turtle.write(arg): Write the arg object on the screen, used in this case to write "year" "month" "day" and ":".

(8)turtle.setup(): Set up the main window size and position.

(9)turtle.speed(speed=None): Set the movement speed during drawing. Speed can range from 0 to 10, speed gradually increases from 1 to 10, with 0 being special and representing the fastest speed.

(10)tutle.pensize(width=None): Set the thickness of the pen's lines.

(11)turtle.done(): Indicate the end of drawing and keep the interface from exiting, calling the Tkinter mainloop() function.

第8章 实战演练
Chapter 8　Practical Exercises

完整代码参考如下： | Full reference code is as follows:

```python
import turtle
import time

def drawGap():
    #Lift the pen
    turtle.penup()
    #Move 5 pixels along the current direction to create a rounded corner effect at the contact point
    turtle.fd(5)
def drawLine(draw):
    #Leave a gap first
    drawGap()
    #Decide whether to put the pen down based on the value of draw
    turtle.pendown() if draw else turtle.penup()
    #Move 60 pixels along the current direction
    turtle.fd(60)
    #Create a gap
    drawGap()
    #Rotate the pen direction clockwise by 90 degrees
    turtle.right(90)

def drawDigit(digit):
#Draw solid line for the middle segment if the digit is 2, 3, 4, 5, 6, 8, or 9, otherwise
#move without drawing
    drawLine(True) if digit in(2, 3, 4, 5, 6, 8, 9) else drawLine(False)
    drawLine(True) if digit in(0, 1, 3, 4, 5, 6, 7, 8, 9) else drawLine(False)
    drawLine(True) if digit in(0, 2, 3, 5, 6, 8, 9) else drawLine(False)
    drawLine(True) if digit in(0, 2, 6, 8) else drawLine(False)
    #Rotate the pen direction counterclockwise by 90 degrees
    turtle. left(90)
    drawLine(True) if digit in(0, 4, 5, 6, 8, 9) else drawLine(False)
    drawLine(True) if digit in(0, 2, 3, 5, 6, 7, 8, 9) else drawLine(False)
    drawLine(True) if digit in(0, 1, 2, 3, 4, 7, 8, 9) else drawLine(False)
    #Rotate the pen direction counterclockwise by 180 degrees
    turtle. left(180)
    #Lift the pen
    turtle. penup()
    #Move 20 pixels along the current direction to prepare for drawing the next time element
    turtle. fd(20)

def drawDate(date):
    #date format example: '2024-06=22+10: 52'
    #Set pen color
```

243

```python
            turtle.pencolor('red')
            #Traverse the time string
            for i in date:
                #Write "Year" at "-" position
                if i=='-':
                    turtle.write('Year', font=('Arial', 18, 'normal'))
                    turtle.pencolor('green')
                    turtle.fd(64)
                #Write "Month" at "=" position
                elif i=='=':
                    turtle.write('Month', font=('Arial', 18, 'normal'))
                    turtle.pencolor('blue')
                    turtle.fd(80)
                #Write "Day" at "+" position
                elif i=='+':
                    turtle.write('Day', font=('Arial', 18, 'normal'))
                    turtle.pencolor('orange')
                    turtle.fd(48)
                #Write ":" at ":" position
                elif i==':':
                    turtle.write(':', font=('Arial', 18, 'normal'))
                    turtle.fd(40)
                #Draw the time digits
                else:
                    drawDigit(eval(i))
def main():
    #Set the width and height of the window
    turtle.setup(1600, 300)
    #Set the speed of movement
    turtle.speed(0)
    #Lift the pen
    turtle.penup()
    #Move back 600 pixels along the current direction
    turtle.fd(-600)
    #Set the thickness of the pen line
    turtle.pensize(13)
    #Get the current date and time data
    #Format example: '2024-06=22+10:52'
    date=time.strftime('%Y-%m=%d+%H:%M', time.localtime(time.time()))
    #Draw the seven-segment display
    drawDate(date)
    #Make the turtle invisible
    turtle.hideturtle()
```

```
        #Drawing is done, keep the interface from exiting
        turtle.done()

    main()
```

代码分析。

（1）drawGap()函数用于绘制7条线段之间的间隙，模拟连接处圆角效果，实际上是向前移动5像素单位，不绘制任何内容。

（2）drawLine()函数根据参数draw的值确定落笔还是抬笔，如果draw的值为True，则落笔，移动绘制；否则抬笔，移动不绘制。

（3）drawDigit()函数根据传入的参数digit数值，绘制digit数字，绘制过程为前面介绍的绘制7条线段，根据digit值判断当前绘制的线段是移动绘制，还是只移动不绘制。

（4）drawDate()函数根据传入的当前日期和时间，绘制七段管时间，遍历日期时间字符串每个位置上的字符，如果是"-""=""+"或":"，则将对应的"年""月""日"或":"字符写到屏幕上，否则调用步骤（3）中的drawDigit()函数来绘制七段管数字。

（5）main()函数进行初始化操作，包括设置绘制的窗体大小、设置画笔移动速度、初始化画笔最初绘制位置、设置画笔线条的粗细等。

Code analysis.

(1)The drawGap() function is used to draw the gap between the 7 segments to simulate rounded corners. It actually moves forward by 5 pixels without drawing anything.

(2)The drawLine() function determines whether to put the pen down or lift it based on the value of the parameter draw. If draw is True, the pen is put down and moves to draw; otherwise, the pen is lifted and moves without drawing.

(3)The drawDigit() function draws the digit number based on the input parameter digit. The process involves drawing 7 segments as described earlier, with the current segment being drawn or only moved based on the digit value.

(4)The drawDate() function draws the seven-segment display time based on the current date and time. It iterates over each character in the date and time string, if it is "-" "=" "+" or ":", then write the corresponding "year" "month" "day" or ":" character on the screen. Otherwise, it calls the drawDigit() function from step(3) to draw the seven-segment digit.

(5)The main() function initializes the drawing operations, including setting the window size for drawing, setting the speed of the pen movement, initializing the initial drawing position of the pen, and setting the thickness of the pen's lines.

8.5 使用scipy库进行图像处理
8.5 Using the scipy Library for Image Processing

scipy是建立在numpy之上的一系列数学算法，通过为用户提供

scipy is a collection of mathematical algorithms on top of numpy. By providing advanced commands and

用于操作和可视化数据的高级命令和类，它为 Python 增添了重要功能。scipy 涵盖了不同科学计算领域的子包，主要包括聚类算法、物理和数学中常数、离散傅里叶变换、积分和常微分方程求解、线性代数、插值、统计分析、N 维图像处理、信号处理等功能。

这一小节里，我们将学习使用 scipy 库中 ndimage 模块中的方法对图像进行处理，包括图像灰度处理、图像二值化和图像高斯平滑处理、分别使用 Sobel 算法、Prewitt 算法和 Laplace 算法对图像进行边缘检测，以及图像的侵蚀和膨胀。scipy 的图像模块实质上将图像视为数组进行处理，虽然不是专门的图像处理库，但其处理的速度非常快。

由于 scipy 依赖 numpy，显示图片处理结果需要使用 matplotlib 库，所以需要安装的第三方库有 numpy、scipy 和 matplotlib。安装命令如下：

classes for manipulating and visualizing data, it adds significant capabilities to Python. scipy covers various subpackages of scientific computing domains. These include clustering algorithms, physical and mathematical constants, discrete Fourier transforms, integration and ordinary differential equation solvers, linear algebra, interpolation, statistical analysis, N-dimensional image processing, signal processing, and more.

In this section, we will learn to process images using methods from the ndimage module in the scipy library. This includes image grayscale processing, image binarization processing, Gaussian smoothing processing of images, edge detection using the Sobel algorithm, Prewitt algorithm, and Laplace algorithm, as well as image erosion and dilation. The image module in scipy essentially treats images as arrays for processing. Even though it is not a dedicated image processing library, its processing speed is very fast.

Since scipy depends on numpy, one will need to install the numpy library. To display the processed images, one will also need the matplotlib library. Therefore, the third-party libraries that need to be installed are numpy, scipy, and matplotlib. Below are the installation commands:

```
pip install numpy
pip install matplotlib
pip install scipy
```

完整参考代码如下： The complete reference code is as follows:

```
import numpy as np
import matplotlib.pyplot as plt
from scipy import ndimage

def image_handle(image_path):
    #Read the image file
    image=plt.imread(image_path)
    #Set the font
```

8-5

```python
plt.rcParams['font.family']='SimHei'
#Display 9 images in 3 rows and 3 columns
fig, ax=plt.subplots(3, 3)
#Original image
ax[0][0].set_title("1. Original Image")
ax[0][0].imshow(image)

#Convert the image to grayscale
gray=np.mean(image, axis=-1)
ax[0][1].set_title("2. Grayscale Image")
ax[0][1].imshow(gray, cmap='gray')

#Binarize the image
binary=gray>0.5
ax[0][2].set_title("3. Binarized Image")
ax[0][2].imshow(binary, cmap='gray')

#Gaussian smoothing
edges=ndimage.gaussian_filter(gray, sigma=2)
ax[1][0].set_title("4. Smoothed Image(Gaussian Smoothing)")
ax[1][0].imshow(edges, cmap='gray')

#Edge detection using the Sobel algorithm
edges=ndimage.sobel(gray)
ax[1][1].set_title("5. Edge Detection(Sobel Algorithm)")
ax[1][1].imshow(edges, cmap='gray')

#Edge detection using the Prewitt algorithm
edges=ndimage.prewitt(gray)
ax[1][2].set_title("6. Edge Detection(Prewitt Algorithm)")
ax[1][2].imshow(edges, cmap='gray')

#Edge detection using the Laplace algorithm
edges=ndimage.laplace(gray)
ax[2][0].set_title("7. Edge Detection(Laplace Algorithm)")
ax[2][0].imshow(edges, cmap='gray')

#Return ndarray data type, 3 rows and 3 columns, all values are True
structure=ndimage.generate_binary_structure(2, 2)

#Image erosion
erosion=ndimage.binary_erosion(binary, structure)
ax[2][1].set_title("8. Eroded Image")
```

```
ax[2][1].imshow(erosion, cmap='gray')

#Image dilation
dilation=ndimage.binary_dilation(binary, structure)
ax[2][2].set_title("9. Dilated Image")
ax[2][2].imshow(dilation, cmap='gray')

#Do not display axis data
for i in range(3):
    for j in range(3):
        ax[i][j].axis('off')

#Maximize the window upon display
mng=plt.get_current_fig_manager()
mng.window.showMaximized()
plt.show()

if __name__=='__main__':
    #Path to the image file
    image_path='pic.png'
    image_handle(image_path)
```

程序运行结果如图8-11所示。

The program execution results are shown in Figure 8-11.

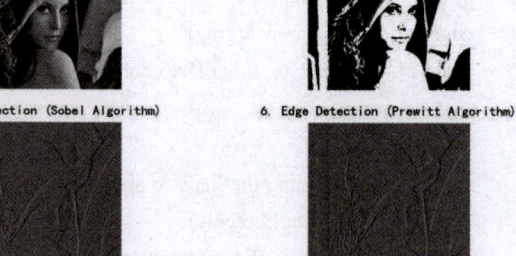

图 8-11 程序运行结果

Figure 8-11 The Program Execution Results

1 是原始图片，通过读取图片文件获取；2 是灰度处理后的图像；3 是二值化后的图像；4 是高斯平滑后的图像，5、6、7 是边缘检测的结果，分别使用的是 Sobel 算法、Prewitt 算法和 Laplace 算法；8 是侵蚀后的图像；9 是膨胀后的图像。

图像灰度化是将一幅彩色图像转换为灰度图像的过程。图像的二值化，就是将图像上的像素点的灰度值设置为 0 或 255，也就是将整个图像呈现出明显的只有黑和白的视觉效果。高斯平滑也称高斯模糊，作用是使图像变得模糊且平滑，以减少图像噪声和降低细节层次。图像边缘检测用于定位图像中的边缘部分。侵蚀和膨胀是图像形态学中最基本的两种操作，用于突出目标的特征。简单来说，侵蚀操作会扩张图像中黑色的区域，膨胀操作会扩张图像中白色的区域。

代码分析。

（1）图片文件读取。使用 matplotlib.pyplot 中的 imread()方法来读取图片文件，读取结果的数据类型是 numpy 中的 ndarray 类型（即由多个具有相同类型和尺寸的元素组成的多维容器），在这里返回的是二维数据，类似于 C 语言中的二维数组。代码为"image = plt.imread(image_path)"。

（2）创建显示容器。使用 matplotlib.pyplot 中的 subplots()方法创建 3 行 3 列的容器，来显示 9 张图片，代码为"fig, ax = plt.subplots(3, 3)"。

（3）图像灰度处理。使用 numpy 中的 mean()方法对图像进行灰度处理，代码为"gray = np.mean(image, axis = -1)"。

1 is the original image obtained by reading the image file; 2 is the image after grayscale processing; 3 is the image after binarization; 4 is the image after Gaussian smoothing; 5 is the result of edge detection using the Sobel algorithm; 6 is the result of edge detection using the Prewitt algorithm; 7 is the result of edge detection using the Laplace algorithm; 8 is the image after erosion; 9 is the image after dilation.

Grayscale conversion in images involves transforming a color image into a grayscale one. Image binarization is the process of setting the grayscale values of pixels in an image to either 0 or 255, resulting in a visual effect where the entire image appears in distinct black and white tones. Gaussian smoothing, also known as Gaussian blur, aims to blur and smooth the image to reduce noise and lower detail levels. Image edge detection is used to locate the edge portions within an image. Erosion and dilation are the two fundamental operations in image morphology, used to enhance the features of the target. In simple terms, the erosion operation expands the black regions in the image, while the dilation operation expands the white regions in the image.

Code analysis.

(1) Image File Reading. Use the imread() method from matplotlib.pyplot to read the image file. The result is a numpy ndarray(a multi-dimensional container consisting of elements of the same type and size), where in this case the returned data is two-dimensional, similar to a 2D array in C. The code is "image = plt.imread(image_path)".

(2) Creating display containers. Use the subplots() method from matplotlib.pyplot to create a 3×3 container for displaying 9 images. The code is "fig, ax = plt.subplots(3, 3)".

(3) Grayscale image processing. Use the mean() method from numpy to process the image into grayscale. The code is "gray = np.mean(image, axis = -1)".

（4）图像二值化处理。对灰度图像二维数据进行筛选，值大于0.5的数据显示为255，即白色；小于0.5的图像显示为0，即黑色。筛选代码为"binary=gray>0.5"。

（5）图像高斯平滑处理。对灰度图像数据进行处理，代码为"ndimage.gaussian_filter(gray, sigma=2)"，其中，sigma可以表示高斯滤波器在每个轴上的标准差值。

（6）边缘检测。边缘检测分别使用了Sobel算法、Prewitt算法和Laplace算法，代码分别为"edges=ndimage.sobel(gray)" "edges=ndimage.prewitt(gray)" "edges=ndimage.laplace(gray)"。

（7）图像的侵蚀和膨胀。图像侵蚀的代码为"erosion=ndimage.binary_erosion(binary, structure)"；图像膨胀的代码为"dilation=ndimage.binary_dilation(binary, structure)"。其中，"structure=ndimage.generate_binary_structure(2, 2)"，structure为二维数据，值均为True。

（8）图像显示。图像显示使用matplotlib.pyplot中的imshow()方法。例如，显示原始图像的代码为"ax[0][0].imshow(image)"。

(4) Image binarization. Process the 2D grayscale image data by setting values greater than 0.5 to 255 (white) and values less than 0.5 to 0(black). The code is "binary=gray>0.5".

(5)Gaussian smoothing. Process the grayscale image data using "ndimage.gaussian_filter(gray, sigma=2)", where sigma represents the standard deviation of the Gaussian filter on each axis.

(6) Edge detection. Different algorithms such as Sobel, Prewitt, and Laplace are used for edge detection. The codes are: "edges=ndimage.sobel(gray)"; "edges=ndimage.prewitt(gray)"; "edges=ndimage.laplace(gray)".

(7)Erosion and dilation. Code for erosion is "erosion=ndimage.binary_erosion(binary, structure)"; code for dilation is "dilation=ndimage.binary_dilation(binary, structure)". Here, "structure=ndimage.generate_binary_structure(2, 2)", structure is a 2D data, where the values are all True.

(8)Image display. Use the imshow() method from matplotlib.pyplot to display images. For example, displaying the original image can be done with "ax[0][0].imshow(image)".

8.6 本章小结
8.6 Summary of This Chapter

本章主要介绍了Python常用第三方库的安装和使用，包括pandas库、matplotlib库、jieba库、wordcloud库、numpy库、scipy库和标准库turtle等，以及这些库中部分方法的使用。Python非常流行的一个重要原因就是拥有众多优秀的第

This chapter mainly introduces the installation and usage of commonly used third-party libraries in Python, including pandas, matplotlib, jieba, wordcloud, numpy, scipy, as well as the standard library turtle. It also covers the usage of some methods in these libraries. Python's popularity is largely attributed to the abundance of excellent third-party libraries. Readers can explore and learn

三方库，读者可以根据需要，通过第三方库的官方网站或帮助文档学习使用第三方库，辅助自己完成软件功能的开发。

how to use these libraries by referring to the official websites or documentation, which will assist in developing software functionalities according to their needs.

8.7 本章练习
8.7 Exercise for This Chapter

8.1 表8-2所示为2024年6月TIOBE网站统计的编程语言排行情况（截取排名前10的数据），请根据该表数据，绘制饼图，展示每种编程语言的流行情况。

8.1 As shown in Table 8-2, the ranking of programming languages in June 2024 on the TIOBE website (selecting the top 10 rankings), please draw a pie chart based on the data in the table to demonstrate the popularity of each programming language.

表8-2 TIOBE 编程语言排行情况（2024年6月）
Table 8-2 TIOBE Programming Language Rankings(Jun-24)

2024年6月 Jun-24	编程语言 Programming Language	占有率 Ratings/%
1	Python	15.39
2	C++	10.03
3	C	9.23
4	Java	8.40
5	C#	6.65
6	JavaScript	3.32
7	Go	1.93
8	SQL	1.75
9	Visual Basic	1.66
10	Fortran	1.53

8.2 使用中文分词库jieba对《西游记》进行分析，统计其中出现频率最高的前50个词汇，并使用wordcloud库绘制词云。

8.3 假设成绩数据格式如表8-1所示，将其存储在Excel文件中，文件名为score.xlsx，使用pandas库读取该Excel文件数据，文件第一行是表头。对读取出的数据进行

8.2 Analyze *Journey to the West* using the Chinese word segmentation library jieba, calculate the top 50 most frequently occurring words, and draw a word cloud using the wordcloud library.

8.3 Assuming the format of the score data is as shown in Table 8-1, stored in an Excel file named score.xlsx, use the pandas library to read the data from this Excel file where the first row represents the header. Perform statistical analysis on the data, including calculating the highest

251

统计，统计出每门课程的最高分、最低分和平均分。

8.4 使用 turtle 库绘制五角星，绘制结果如图 8-12 所示。

score, lowest score, and average score for each course.

8.4 Use the turtle library to draw a pentagon star, the result of the drawing is shown in Figure 8-12.

图 8-12　五角星

Figure 8-12　Pentagon Star

本章练习题参考答案

参 考 文 献

[1] 汪彦，何骞，胡奇光，等. Python 语言程序设计[M]. 北京：北京邮电大学出版社，2024.
[2] 朱大勇，陈佳，许毅. Python 程序设计基础与应用[M]. 北京：人民邮电出版社，2023.
[3] 戴凤智，程宇辉，冀承绪. Python 入门边学边练[M]. 北京：化学工业出版社，2024.
[4] 赵广辉，李屾，秦珀石，等. Python 程序设计基础实践教程[M]. 北京：高等教育出版社，2021.
[5] 刘德山，杨洪伟，崔晓松. Python 3 程序设计[M]. 第 2 版. 北京：人民邮电出版社，2022.
[6] 林子雨，赵江声，陶继平. Python 程序设计基础教程[M]. 北京：人民邮电出版社，2022.
[7] 嵩天，礼欣，黄天宇. Python 语言程序设计基础[M]. 第 2 版. 北京：高等教育出版社，2017.
[8] BEAZLEY D M. Python 参考手册[M]. 第 2 版. 谢俊，杨越，高伟，译. 北京：人民邮电出版社，2016.
[9] CUNNINGHAM K. Python 入门经典[M]. 李军，李强，译. 北京：人民邮电出版社，2014.